Teacher Resources

Complete Solutions Guide

to accompany

World of Chemistry

Steven S. Zumdahl
Susan L. Zumdahl
Donald J. DeCoste

McDougal Littell
A Houghton Mifflin Company

Evanston, IL
Boston • New York

Senior Sponsoring Editor: Richard Stratton
Editorial Associate: Karen O'Connor
Editorial Assistant: Marisa R. Papile
Senior Project Editor: Tamela Ambush
Senior Manufacturing Coordinator: Florence Cadran
Executive Marketing Manager: Andy Fisher

Copyright © 2002 by Houghton Mifflin Company. All rights reserved.

Houghton Mifflin Company hereby grants you permission to reproduce the Houghton Mifflin material contained in this work in classroom quantities, solely for use with the accompanying Houghton Mifflin textbook. All reproductions must include the Houghton Mifflin copyright notice, and no fee may be collected except to cover the cost of duplication. If you wish to make any other use of this material, including reproducing or transmitting the material or portions thereof in any form or by any electronic or mechanical means including any information storage or retrieval system, you must obtain prior written permission from Houghton Mifflin Company, unless such use is expressly permitted by federal copyright law. If you wish to reproduce material acknowledging a rights holder other than Houghton Mifflin, you must obtain permission from the rights holder. Address inquires to College Permissions Houghton Mifflin Company, 222 Berkeley Street, Boston, MA 02116-3764.

Printed in the U.S.A
ISBN: 0-618-07229-2

5 6 7 8 9-HS-05

Contents

	Preface	v
Chapter 1	Chemistry: An Introduction	1
Chapter 2	Matter	3
Chapter 3	Chemical Foundations: Elements, Atoms, and Ions	6
Chapter 4	Nomenclature	14
Chapter 5	Measurements and Calculations	26
Chapter 6	Chemical Composition	37
Chapter 7	Chemical Reactions: An Introduction	58
Chapter 8	Reactions in Aqueous Solutions	63
Chapter 9	Chemical Quantities	75
Chapter 10	Energy	99
Chapter 11	Modern Atomic Theory	108
Chapter 12	Chemical Bonding	117
Chapter 13	Gases	128
Chapter 14	Liquids and Solids	143
Chapter 15	Solutions	151
Chapter 16	Acids and Bases	165
Chapter 17	Equilibrium	176
Chapter 18	Oxidation-Reduction Reactions /Electrochemistry	186
Chapter 19	Radioactivity and Nuclear Energy	198
Chapter 20	Organic Chemistry	205
Chapter 21	Biochemistry	219

Preface

This guide contains complete solutions for the end-of-chapter problems in *World of Chemistry*. We have tried to give the most detailed solutions possible to all the problems, and we have made a conscious effort to solve each problem in the manner discussed in the textbook. You may, of course, wish to discuss alternative methods of solution with your students.

One topic that causes many students concern is the matter of significant figures, and the determination of the number of digits to which a solution to a problem should be reported. To avoid truncation errors, the solutions typically report intermediate answers to one more digit than appropriate for the final answer. The final answer to each problem is then given to the correct number of significant figures based on the data provided in the problem.

We hope you find the Complete Solutions Guide useful. Have a great year!

S.S.Z.
S.L.Z.
D.J.D.

Chapter 1 Chemistry: An Introduction

1. Obviously, the answer to this question depends on your own experience. You might consider such things as how oven and drain cleaners work, why antifreeze keeps your car's radiator from freezing, why cuts and scrapes are often cleaned with hydrogen peroxide, how a "permanent wave" curls hair, etc.

2. Examples: physician (understanding cellular processes, understanding how drugs and blood tests work); lawyer (understanding scientific/forensic laboratory tests for use in court); pharmacist (understanding how drugs work, and interactions between drugs); artist (understanding the various media used in art work); photographer (understanding how the film exposure and developing chemical processes occur and how they can be controlled and modified); farmer (understanding which pesticide or fertilizer is needed and how these chemicals work); nurse (understanding how various tests and drugs may affect the patient's well-being).

3. There are obviously many such examples. Many new drugs and treatments have recently become available thanks to research in biochemistry and cell biology. New long-wearing, more comfortable contact lenses have been produced by research in polymer and plastics chemistry. Special plastics and metals were prepared for the production of compact discs to replace vinyl phonograph records.

4. There are, unfortunately, many examples. Chemical and biological weapons are still being produced in some countries. Although the development of new plastics has been a boon in many endeavors, this also increases our depletion of fossil fuels and our solid waste problems. Although many exciting new drugs and treatments have become available, the same biotechnology may lead to testing procedures for determining whether a person has a genetic likelihood of developing a particular disease, which may make it impossible or difficult for that person to obtain health or life insurance.

5. This answer depends on your own experience.

6. This answer depends on your own experience, but consider the following examples: oven cleaner (the label says it contains sodium hydroxide; it converts the burned-on grease in the oven to a soapy material that washes away); drain cleaner (the label says it contains sodium hydroxide; it dissolves the clog of hair in the drain); stomach antacid (the label says it contains calcium carbonate; it makes me belch and makes my stomach feel better); hydrogen peroxide (the label says it is a 3% solution of hydrogen peroxide; when applied to a wound, it bubbles); depilatory cream (the label says it contains sodium hydroxide; it removes unwanted hair from skin).

7. Answer will depend on student experience.

8. David and Susan first recognized the problem (unexplained medical problems). A possible explanation was then proposed (the glaze on their china might be causing lead poisoning). The explanation was tested by experiment (it was determined that the china did contain lead). A full discussion of this scenario is given in the text.

9. The steps are: (1) recognizing the problem and stating it clearly; (2) proposing possible solutions to or explanations of the problem; and (3) performing experiments to test the solutions or explanations.

10. a. quantitative - a number (measurement) is indicated explicitly
    b. qualitative - only a qualitative description is given
    c. quantitative - a numerical measurement is indicated
    d. qualitative - only a qualitative description is given
    e. quantitative - a number (measurement) is implied
    f. qualitative - a qualitative judgment is given
    g. quantitative - a numerical quantity is indicated.

11. A hypothesis is a *possible* explanation of a single observed phenomenon. A theory or model consists of a set of tested hypotheses which give an overall explanation of some part of nature.

12. A natural law is a *summary of observed, measurable behavior* that occurs repeatedly and consistently. A theory is our attempt to *explain* such behavior. The conservation of mass observed during chemical reactions is an example of a natural law. The idea that the universe began with a "big bang" is an example of a theory.

13. Scientists are human, too. When a scientist formulates a hypothesis, he or she wants it to be proven correct. In academic research, for example, scientists want to be able to publish papers on their work to gain renown and acceptance from their colleagues. In industrial situations, the financial success of the individual and of the company as a whole may be at stake. Politically, scientists may be under pressure from the government to "beat the other guy."

14. Most applications of chemistry are oriented toward the interpretation of observations and the solving of problems. Although memorization of some facts may *aid* in these endeavors, it is the ability to combine, relate, and synthesize information that is most important in the study of chemistry.

15. Chemistry is not merely a list of observations, definitions, and properties. Chemistry is the study of very real interactions among different samples of matter, whether within a living cell, or in a chemical factory. When we study chemistry, at least in the beginning, we try to be as general and as nonspecific as possible, so that the *basic principles* learned can be applied to many situations. In a beginning chemistry course, we learn to interpret and solve a basic set of very simple problems, in the hopes that the method of solving these simple problems can be extended to more complex real life situations later on. The actual solution to a problem, at this point, is not as important as learning how to recognize and interpret the problem, and how to propose reasonable, experimentally testable hypotheses.

16. In real life situations, the problems and applications likely to be encountered are not simple textbook examples. One must be able to observe an event, hypothesize a cause, and then test this hypothesis. One must be able to carry what has been learned in class forward to new, different situations.

Chapter 2 Matter

1. We can "see" atoms with a scanning tunneling microscope.

2. Atoms are very tiny.

3. When we say that a compound always has the same composition, we mean that the molecules of that compound always contain the same type and number of atoms of its constituent elements.

4. Typically, the properties of a compound and the elements that constitute it are very different. Consider the properties of liquid *water* and the hydrogen and oxygen gases from which the water was prepared. Consider the properties of *sodium chloride* (table salt) and the sodium metal and chlorine gas from which it might have been prepared.

5. The gaseous 10-g sample of water has a much *larger* volume than either the solid or liquid samples. While the 10-g sample of water vapor contains the same amount of water as the solid and liquid sample (the same number of water molecules), there is a great deal of empty space in the gaseous sample.

6. Gases are easily compressed into smaller volumes, whereas solids and liquids are not. Since a gaseous sample consists mostly of empty space, it is this empty space which is compressed when pressure is applied to a gas.

7. physical

8. chemical

9. the orange color

10. the substance reacts with iron(II) sulfate

11. Electrolysis is the passage of an electrical current through a substance or solution to force a chemical reaction to occur. Electrolysis causes chemical changes to take place that would ordinarily not take place on their own. When an electrical current is passed through water, the current causes the water molecules to break down into their constituent elements (hydrogen gas and oxygen gas).

12.
   a. chemical; the scorch represents the oxidation of the material
   b. physical; the gas in the tires decreases in volume with temperature
   c. chemical; tarnish on silver is caused by reaction of the silver with sulfur or oxygen
   d. chemical; the oven cleaner contains sodium hydroxide which converts greases in the oven into soaps
   e. chemical; an ordinary flashlight battery is constructed with a zinc casing which serves as one of the electrodes. As the battery discharges, the zinc is oxidized.
   f. chemical; the acids attack the calcium phosphate matrix of the teeth

g. chemical; the charring represents the breakdown of the sugar

h. chemical; iron in the blood catalyzes the decomposition of the hydrogen peroxide into oxygen gas (and water)

i. physical; this is just a change in state; carbon dioxide, only in the gaseous state, is still present after the sublimation

j. chemical; chlorine is an oxidizing agent and can change the chemical nature of the dyes in the fabrics

13. solutions: window cleaner, shampoo, rubbing alcohol
    mixtures: salad dressing, jelly beans, the change in my pocket

14. a. mixture
    b. mixture
    c. mixture
    d. pure substance

15. a. homogeneous
    b. heterogeneous
    c. heterogeneous
    d. homogeneous
    e. the paper itself is basically homogeneous in appearance

16. Consider a salt solution (sodium chloride in water). Since water boils at a much lower temperature than sodium chloride, the water can be boiled off from the solution, collected, and subsequently condensed back into the liquid state. This separates the two chemical substances.

17. Consider a mixture of salt (sodium chloride) and sand. Salt is soluble in water, sand is not. The mixture is added to water and stirred to dissolve the salt, and is then filtered. The salt solution passes through the filter, the sand remains on the filter. The water can then be evaporated from the salt.

18. If water is added to the sample, and the sample is then heated to boiling, this should dissolve the benzoic acid but not the charcoal. The hot sample could then be *filtered,* which would remove the charcoal. The solution which passed through the filter could then be cooled, which should cause some of the benzoic acid to crystallize, or the solution could be heated carefully to boil off the water leaving benzoic acid behind.

19. The solution is heated to vaporize (boil) the water. The water vapor is then cooled so that it condenses back to the liquid state, and the liquid is collected. After all the water is vaporized from the original sample, pure sodium chloride will remain. The process consists of physical changes.

20. Since X is a pure substance, the fact that two different solids form when electrical current is passed indicates that X must be a compound.

Matter

21. Chalk must be a compound, since it loses mass when heated, and appears to change into a substance with different physical properties (the hard chalk turns into a crumbly substance).

22. Since vaporized water is still the *same substance* as solid water, no chemical reaction has occurred. Sublimation is a physical change.

23. Liquids and gases both flow freely and take on the shape of their container. The molecules in liquids are relatively close together and interact with each other, whereas the molecules in gases are far apart from each other and do not interact with each other.

24. far apart

25. physical

26. chemical

Chapter 3 Chemical Foundations: Elements, Atoms and Ions

1. Although the number and nature of the elementary substances postulated by the ancient Greeks were incorrect, their idea that the matter we encounter in everyday life is composed of a few *simpler* substances is very similar to our modern concepts. Also, the idea that the simpler substances *combine* with each other in regular, fixed manners compares well with the modern theory of matter.

2. The alchemists discovered several previously unknown elements (mercury, sulfur, antimony) and were the first to prepare several common acids.

3. Boyle's most important contribution was his insistence that science should be firmly grounded in *experiment*. Boyle tried to limit the influence of any preconceptions about science, and only accepted as fact what could be demonstrated.

4. There are 112 elements presently known; of these 88 occur naturally and 24 are manmade.

5. Oxygen is found in great abundance in the oceans (combined with hydrogen in water molecules) and in the earth itself (most rocks and minerals are oxygen compounds). Oxygen is found more commonly in compounds.

6. The four most abundant elements in living creatures are, respectively, oxygen, carbon, hydrogen, and nitrogen. In the nonliving world, the most abundant elements are, respectively, oxygen, silicon, aluminum, and iron.

7. B (boron)
   C (carbon)
   F (fluorine)
   H (hydrogen)
   I (iodine)
   K (potassium)
   N (nitrogen)
   O (oxygen)
   P (phosphorus)
   S (sulfur)
   U (uranium)
   V (vanadium)
   W (tungsten)
   Y (yttrium)

8. Sb (antimony)
   Cu (copper)

Chemical Foundations: Elements, Atoms, and Ions

Au (gold)

Pb (lead)

Hg (mercury)

K (potassium)

Ag (silver)

Na (sodium)

Sn (tin)

W (tungsten)

Fe (iron)

9.
   a. Ne
   b. Ni
   c. K
   d. Si
   e. Ba
   f. Agi

10.

Symbol	Name
Fe	iron
Cl	chlorine
S	sulfur
U	uranium
Ne	neon
K	potassium

11.
   a. copper
   b. cobalt
   c. calcium
   d. carbon
   e. chromium
   f. cesium
   g. chlorine
   h. cadmium

12.
   a. False; most materials occur as mixtures of compounds.
   b. False; a given compound *always* contains the same relative number of atoms of its various elements.

c. False; molecules are made up of tiny particles called atoms.

13. According to Dalton, a given compound is always made up of the same number and type of atoms, and so the composition of the compound on a mass percentage basis will always be the same, no matter what the source of the compound is. For example, water molecules from the Atlantic Ocean and the Pacific Ocean all contain two hydrogen atoms bonded to one oxygen atom.

14. A compound is a distinct substance that is composed of two or more elements and always contains exactly the same relative masses of those elements.

15. According to Dalton, all atoms of the same element are *identical*; in particular, every atom of a given element has the same *mass* as every other atom of that element. If a given compound always contains the *same relative numbers* of atoms of each kind, and those atoms always have the *same masses,* then it follows that the compound made from those elements would always contain the same relative masses of its elements.

16.
   a. PCl_3
   b. B_2H_6
   c. $CaCl_2$
   d. CBr_4
   e. Fe_2O_3
   f. H_3PO_4

17.
   a. False; Thomson obtained beams of *identical* particles whose whose nature did *not* depend on what gas was used to generate them.
   b. True
   c. False; the atom was envisioned as a sphere of *positive* charge in which *negatively* charged electrons were randomly distributed.

18.
   a. False; Rutherford's bombardment experiments with metal foil suggested that the alpha particles were being deflected by coming near a *dense, positively charged* atomic nucleus.
   b. False; The proton and the electron have opposite charges, but the mass of the electron is *much smaller* than the mass of the proton.
   c. True

19. neutrons

20. protons

21. The proton and the neutron have similar (but not identical) masses. Either of these particles has a mass approximately 2000 times greater than that of an electron. The combination of the protons and the neutrons make up the bulk of the mass of an atom, but the electrons make the greatest contribution to the chemical properties of the atom.

Chemical Foundations: Elements, Atoms, and Ions

22. electrons

23. False; atoms that have the same number of *protons*, with different numbers of *neutrons*, represent isotopes.

24. False; the mass number represents the total number of protons and neutrons in the nucleus.

25. Dalton's original theory proposed that all atoms of a given element were *identical*. We now realize that different atoms of the same element must have a particular number of protons and electrons (the atomic number), but may have different numbers of neutrons (leading to different mass numbers).

26. Atoms of the same element (i.e., atoms with the same number of protons in the nucleus) may have different numbers of neutrons, and so will have different masses.

27.
   a. 32
   b. 30
   c. 24
   d. 74
   e. 38
   f. 27
   g. 4
   h. 3

28.
   a. $^{17}_{8}O$
   b. $^{37}_{17}Cl$
   c. $^{60}_{27}Co$
   d. $^{57}_{26}Fe$
   e. $^{131}_{53}I$
   f. $^{7}_{3}Li$

29.
   a. 94 protons, 150 neutrons, 94 electrons
   b. 95 protons, 146 neutrons, 95 electrons
   c. 89 protons, 138 neutrons, 89 electrons
   d. 55 protons, 78 neutrons, 55 electrons

Copyright © Houghton Mifflin Company. All rights reserved.

e. 77 protons, 116 neutrons, 77 electrons

f. 25 protons, 31 neutrons, 25 electrons

30.

element	symbol	atomic number	mass number	number of neutrons
sodium	$^{23}_{11}Na$	11	23	12
nitrogen	$^{15}_{7}N$	7	15	8
barium	$^{136}_{56}Ba$	56	136	80
lithium	$^{9}_{3}Li$	3	9	6
boron	$^{11}_{5}B$	5	11	6

31. The elements are listed in the periodic table in order of increasing atomic number (number of protons in the nucleus). The periodic table originally was arranged on the basis of mass.

32. Elements with similar chemical properties are aligned *vertically* in families known as *groups*.

33. Metals are excellent conductors of heat and electricity, and are malleable, ductile, and generally shiny (lustrous) when a fresh surface is exposed.

34. Metallic elements are found towards the *left* and *bottom* of the periodic table; there are far more metallic elements than there are nonmetals.

35. Mercury is a liquid at room temperature.

36. hydrogen, nitrogen, oxygen, fluorine, chlorine, plus all the group 8 elements (noble gases)

37. The only metal which ordinarily occurs as a liquid is mercury. The only nonmetallic element which occurs as a liquid at room temperature is bromine (elements such as oxygen and nitrogen are frequently obtainable as liquids, but these result from compression of the gases into cylinders at very low temperatures).

38. The metalloids are the elements found on either side of the "stairstep" region that is marked on most periodic tables. The metalloid elements show some properties of both metals and nonmetals.

39.
a. Group 7; halogens

b. Group 2; alkaline earth elements

c. Group 1; alkali metals

d. Group 1; alkali metals

e. Group 8; noble gases

Chemical Foundations: Elements, Atoms, and Ions

    f.   Group 1; alkali metals

    g.   Group 8; noble gases

40.

name	symbol	atomic number	group number	metal/nonmetal
rubidium	Rb	37	1	metal
germanium	Ge	32	4	metalloid
magnesium	Mg	12	2	metal
titanium	Ti	22	-	transition metal
iodine	I	53	7	nonmetal

41.   Most of the elements are too reactive to be found in the uncombined form in nature, and are found only in compounds.

42.   These elements are found *uncombined* in nature and do not readily react with other elements. For many years it was thought that these elements formed no compounds at all, although this has now been shown to be untrue.

43.   Diatomic gases: H_2, N_2, O_2, Cl_2, and F_2

    Monatomic gases: He, Ne, Kr, Xe, Rn, and Ar

44.   liquids: bromine, mercury, gallium

    gases: hydrogen, nitrogen, oxygen, fluorine, chlorine, and the noble gases (helium, neon, argon, krypton, xenon, radon)

45.
- [1] b
- [2] d
- [3] b
- [4] h
- [5] f
- [6] e
- [7] a
- [8] c
- [9] g
- [10] i

46.

a.	Co^{2+}:	27 protons	25 electrons	CoO
b.	Co^{3+}:	27 protons	24 electrons	Co_2O_3
c.	Cl^-:	17 protons	18 electrons	$CaCl_2$
d.	K^+:	19 protons	18 electrons	K_2O
e.	S^{2-}:	16 protons	18 electrons	CaS
f.	Sr^{2+}:	38 protons	36 electrons	SrO
g.	Al^{3+}:	13 protons	10 electrons	Al_2O_3
h.	P^{3-}:	15 protons	18 electrons	Ca_3P_2

47.
    a.   Ca: 20 protons, 20 electrons Ca^{2+}: 20 protons, 18 electrons
    b.   P: 15 protons, 15 electrons P^{3-}: 15 protons, 18 electrons
    c.   Br: 35 protons, 35 electrons Br^-: 35 protons, 36 electrons
    d.   Fe: 26 protons, 26 electrons Fe^{3+}: 26 protons, 23 electrons
    e.   Al: 13 protons, 13 electrons Al^{3+}: 13 protons, 10 electrons
    f.   N: 7 protons, 7 electrons N^{3-}: 7 protons, 10 electrons

48.
    a. I^-
    b. Sr^{2+}
    c. Cs^+
    d. Ra^{2+}
    e. F^-
    f. Al^{3+}

49.   A compound which has a high melting point (many hundreds of degrees) and which conducts an electrical current when melted or dissolved in water almost certainly consists of ions.

50.   In the solid state, although ions are present, they are rigidly held in fixed positions in the crystal of the substance. In order for ionic substances to be able to pass an electrical current, the ions must be able to *move,* which is possible when the solid is converted to the liquid state.

51.   The total number of positive charges must equal the total number of negative charges so that there will be *no net charge* on the crystals of an ionic compound. A macroscopic sample of compound must ordinarily not have any net charge.

52.
    a.   One 3– ion is needed to balance one 3^+ ion: FeP
    b.   The smallest common multiple of 3 and 2 is 6; three 2– ions are required to balance two 3+ ions: Fe_2S_3
    c.   Three 1– ions are required to balance one 3+ ion: $FeCl_3$
    d.   Two 1– ions are required to balance one 2+ ion: $MgCl_2$
    e.   One 2– ion balances one 2+ ion: MgO
    f.   The smallest common multiple of 2 and 3 is 6; two 3– ions are required to balance three 2+ ions: Mg_3N_2
    g.   Three 1+ ions are required to balance one 3– ion: Na_3P
    h.   Two 1+ ions are required to balance one 2– ion: Na_2S

53.   The atomic number represents the number of protons in the nucleus of an atom. The mass number represents the total number of protons and neutrons. No two different elements have the same atomic number. If the *total* number of protons and neutrons happens to be the same for two atoms then the atoms will have the same mass number.

Chemical Foundations: Elements, Atoms, and Ions

54. Most of the mass of an atom is concentrated in the nucleus: the *protons* and *neutrons* which constitute the nucleus have similar masses, and these particles are nearly two thousand times heavier than electrons. The chemical properties of an atom depend on the number and location of the *electrons* it possesses. Electrons are found in the outer regions of the atom, and are the particles most likely to be involved in interactions between atoms.

55. Yes. For example, carbon and oxygen form carbon monoxide (CO) and carbon dioxide (CO_2). The existence of more than one compound between the same elements does not in any way contradict Dalton's theory. For example, the relative mass of carbon in different samples of CO is always the same, and the relative mass of carbon in different samples of CO_2 is also always the same. Dalton did not say, however, that two different compounds would have to have the same relative masses of the elements present. In fact, Dalton said that two different compounds of the same elements would have to have different relative masses of the elements.

56.
   a. CO_2
   b. $AlCl_3$
   c. $HClO_4$
   d. SCl_6

Copyright © Houghton Mifflin Company. All rights reserved.

Chapter 4   Nomenclature

1. A *binary* compound is one that contains only two elements. Examples are sodium chloride, water, and carbon dioxide.

2. compounds that contain a metal and a nonmetal; compounds containing two nonmetals

3. cation, anion

4. cation

5. The substance "sodium chloride" consists of an extended lattice array of sodium ions, Na^+, and chloride ions, Cl^-. Each sodium ion is surrounded by several chloride ions, and each chloride ion is surrounded by several sodium ions. We write the formula as NaCl to indicate the relative number of each ion in the substance.

6.
   a. sodium iodide
   b. calcium fluoride
   c. aluminum sulfide
   d. calcium bromide
   e. strontium oxide
   f. silver chloride [silver(I) chloride]
   g. cesium iodide
   h. lithium oxide

7.
   a. incorrect; BaH_2 is barium hydride
   b. incorrect; Na_2O is sodium oxide
   c. correct
   d. incorrect; SiO_2 is silicon dioxide
   e. correct

8.
   a. Since each iodide ion has a 1– charge, the iron ion must have a 3+ charge: the name is iron(III) iodide.
   b. Since each chloride ion has a 1– charge, the manganese must have a 2+ charge: the name is manganese(II) chloride.
   c. Since the oxide ion has a 2– charge, the mercury ion must have a 2+ charge: mercury(II) oxide.
   d. Since the oxide ion has a 2– charge, the copper atoms must each have a 1+ charge: the name is copper(I) oxide.
   e. Since the oxide ion has a 2– charge, the copper ion must have a 2+ charge: copper(II) oxide.
   f. Since each bromide ion has a 1– charge, the tin ion must have a 4+ charge: tin(IV) bromide.

9. a. Since each chloride ion has a 1– charge, the cobalt ion must have a 2+ charge: cobalt*ous* chloride.

b. Since each bromide ion has a 1– charge, the chromium ion must have a 3+ charge: the name is chrom*ic* bromide.

c. Since the oxide ion has a 2– charge, the lead ion must have a 2+ charge: the name is plumb*ous* oxide.

d. Since each oxide ion has a 2– charge, the tin ion must have a 4+ charge: the name is stann*ic* oxide.

e. Since each oxide ion has a 2– charge, the iron ion must have a 3+ charge: the name is fer*ric* oxide.

f. Since each chloride ion has a 1– charge, the iron ion must have a 3+ charge: the name is fer*ric* chloride.

10. Remember that for this type of compound of nonmetals, numerical prefixes are used to indicate how many of each type of atom is present. However, if only one atom of the first element mentioned in the compound is present in a molecule, the prefix *mono-* is not needed.

a. iodine pentafluoride

b. arsenic trichloride

c. selenium monoxide

d. xenon tetrafluoride

e. nitrogen triiodide

f. diboron trioxide

11. a. germanium tetrahydride

b. dinitrogen tetrabromide

c. diphosphorus pentasulfide

d. selenium dioxide

e. ammonia (nitrogen trihydride)

f. silicon dioxide

12. a. diboron hexahydride – nonionic (common name: *diborane*)

b. calcium nitride – ionic

c. carbon tetrabromide – nonionic

d. silver sulfide – ionic

e. copper(II) chloride, cupric chloride – ionic

f. chlorine monofluoride – nonionic

13. a. radium chloride – ionic

b. selenium dichloride – nonionic

c. phosphorus trichloride – nonionic
d. sodium phosphide – ionic
e. manganese(II) fluoride – ionic
f. zinc oxide – ionic

14. An oxyanion is a polyatomic anion containing oxygen combined with another element. The following oxyanions of bromine illustrate the nomenclature:

   BrO^- hypobromite
   BrO_2^- bromite
   BrO_3^- bromate
   BrO_4^- perbromate

15. perchlorate, ClO_4^-

16. hypobromite

    IO_3^-

    periodate

    OI^- or IO^-

17. a. NO_3^-
    b. NO_2^-
    c. NH_4^+
    d. CN^-

18. a. Cl^-
    b. ClO^-
    c. ClO_3^-
    d. ClO_4^-

19. a. $MgCl_2$
    b. $Ca(ClO)_2$
    c. $KClO_3$
    d. $Ba(ClO_4)_2$

20. a. permanganate
    b. peroxide
    c. chromate
    d. dichromate
    e. nitrate
    f. sulfite

Nomenclature

21.
   a. iron(III) nitrate, ferric nitrate
   b. cobalt(II) phosphate, cobaltous phosphate
   c. chromium(III) cyanide, chromic cyanide
   d. aluminum sulfate
   e. chromium(II) acetate, chromous acetate
   f. ammonium sulfite

22. An acid is a substance which produces hydrogen ions, H^+, when dissolved in water.

23. oxygen (commonly referred to as *oxy*acids)

24.
   a. hydrochloric acid
   b. sulfuric acid
   c. nitric acid
   d. hydroiodic acid
   e. nitrous acid
   f. chloric acid
   g. hydrobromic acid
   h. hydrofluoric acid
   i. acetic acid

25.
   a. Li_2O
   b. AlI_3
   c. Ag_2O
   d. K_3N
   e. Ca_3P_2
   f. MgF_2
   g. Na_2S
   h. BaH_2

26.
   a. CO_2
   b. SO_2
   c. N_2Cl_4
   d. CI_4
   e. PF_5
   f. P_2O_5

27.
- a. $Ca_3(PO_4)_2$
- b. NH_4NO_3
- c. $Al(HSO_4)_3$
- d. $BaSO_4$
- e. $Fe(NO_3)_3$
- f. $CuOH$

28.
- a. HCN
- b. HNO_3
- c. H_2SO_4
- d. H_3PO_4
- e. $HClO$ or $HOCl$
- f. HBr
- g. $HBrO_2$
- h. HF

29.
- a. $LiCl$
- b. Cu_2CO_3
- c. HBr
- d. $Ca(NO_3)_2$
- e. $NaClO_4$
- f. $Al(OH)_3$
- g. $Ba(HCO_3)_2$
- h. $FeSO_4$
- i. B_2Cl_6
- j. PBr_5
- k. K_2SO_3
- l. $Ba(C_2H_3O_2)_2$

30. A moist paste of NaCl would contain Na^+ and Cl^- ions in solution, and would serve as a *conductor* of electrical impulses.

31. A *binary* compound is a compound containing two and only two elements. A *polyatomic* anion is several atoms bonded together which, as a whole, carries a negative electrical charge. An *oxyanion* is a negative ion containing a particular element and one or more oxygen atoms.

32.
- a. gold(III) bromide, auric bromide
- b. cobalt(III) cyanide, cobaltic cyanide

Nomenclature

    c.   magnesium hydrogen phosphate

    d.   diboron hexahydride (diborane is its common name)

    e.   ammonia

    f.   silver(I) sulfate (usually called silver sulfate)

    g.   beryllium hydroxide

33.   a.   ammonium carbonate

    b.   ammonium hydrogen carbonate, ammonium bicarbonate

    c.   calcium phosphate

    d.   sulfurous acid

    e.   manganese(IV) oxide

    f.   iodic acid

    g.   potassium hydride

34.   a. K_2O

    b. MgO

    c. FeO

    d. Fe_2O_3

    e. ZnO

    f. PbO

    g. Al_2O_3

35. M^+ compounds: MD, M_2E, M_3F

    M^{2+} compounds: MD_2, ME, M_3F_2

    M^{3+} compounds: MD_3, M_2E_3, MF

36. Fe^{2+}:

Formula	Name
$FeCO_3$	iron(II) carbonate; ferrous carbonate
$Fe(BrO_3)_2$	iron(II) bromate; ferrous bromate
$Fe(C_2H_3O_2)_2$	iron(II) acetate; ferrous acetate
$Fe(OH)_2$	iron(II) hydroxide; ferrous hydroxide
$Fe(HCO_3)_2$	iron(II) bicarbonate; ferrous bicarbonate
$Fe_3(PO_4)_2$	iron(II) phosphate; ferrous phosphate
$FeSO_3$	iron(II) sulfite; ferrous sulfite
$Fe(ClO_4)_2$	iron(II) perchlorate; ferrous perchlorate
$FeSO_4$	iron(II) sulfate; ferrous sulfate
FeO	iron(II) oxide; ferrous oxide
$FeCl_2$	iron(II) chloride; ferrous chloride

    Al^{3+}:

Formula	Name
$Al_2(CO_3)_3$	aluminum carbonate
$Al(BrO_3)_3$	aluminum bromate
$Al(C_2H_3O_2)_3$	aluminum acetate
$Al(OH)_3$	aluminum hydroxide

	Al(HCO$_3$)$_3$	aluminum bicarbonate
	AlPO$_4$	aluminum phosphate
	Al$_2$(SO$_3$)$_3$	aluminum sulfite
	Al(ClO$_4$)$_3$	aluminum perchlorate
	Al$_2$(SO$_4$)$_3$	aluminum sulfate
	Al$_2$O$_3$	aluminum oxide
	AlCl$_3$	aluminum chloride
Na$^+$:	Na$_2$CO$_3$	sodium carbonate
	NaBrO$_3$	sodium bromate
	NaC$_2$H$_3$O$_2$	sodium acetate
	NaOH	sodium hydroxide
	NaHCO$_3$	sodium bicarbonate
	Na$_3$PO$_4$	sodium phosphate
	Na$_2$SO$_3$	sodium sulfite
	NaClO$_4$	sodium perchlorate
	Na$_2$SO$_4$	sodium sulfate
	Na$_2$O	sodium oxide
	NaCl	sodium chloride
Ca^{2+}:	CaCO$_3$	calcium carbonate
	Ca(BrO$_3$)$_2$	calcium bromate
	Ca(C$_2$H$_3$O$_2$)$_2$	calcium acetate
	Ca(OH)$_2$	calcium hydroxide
	Ca(HCO$_3$)$_2$	calcium bicarbonate
	Ca$_3$(PO$_4$)$_2$	calcium phosphate
	CaSO$_3$	calcium sulfite
	Ca(ClO$_4$)$_2$	calcium perchlorate
	CaSO$_4$	calcium sulfate
	CaO	calcium oxide
	CaCl$_2$	calcium chloride
NH$_4^+$:	(NH$_4$)$_2$CO$_3$	ammonium carbonate
	NH$_4$BrO$_3$	ammonium bromate
	NH$_4$C$_2$H$_3$O$_2$	ammonium acetate
	NH$_4$OH	ammonium hydroxide
	NH$_4$HCO$_3$	ammonium bicarbonate
	(NH$_4$)$_3$PO$_4$	ammonium phosphate
	(NH$_4$)$_2$SO$_3$	ammonium sulfite
	NH$_4$ClO$_4$	ammonium perchlorate
	(NH$_4$)$_2$SO$_4$	ammonium sulfate
	(NH$_4$)$_2$O	ammonium oxide
	NH$_4$Cl	ammonium chloride
Fe^{3+}:	Fe$_2$(CO$_3$)$_3$	iron(III) carbonate
	Fe(BrO$_3$)$_3$	iron(III) bromate
	Fe(C$_2$H$_3$O$_2$)$_3$	iron(III) acetate
	Fe(OH)$_3$	iron(III) hydroxide
	Fe(HCO$_3$)$_3$	iron(III) bicarbonate
	FePO$_4$	iron(III) phosphate
	Fe$_2$(SO$_3$)$_3$	iron(III) sulfite

Nomenclature

	Fe(ClO$_4$)$_3$	iron(III) perchlorate
	Fe$_2$(SO$_4$)$_3$	iron(III) sulfate
	Fe$_2$O$_3$	iron(III) oxide
	FeCl$_3$	iron(III) chloride

Ni^{2+}:	NiCO$_3$	nickel(II) carbonate
	Ni(BrO$_3$)$_2$	nickel(II) bromate
	Ni(C$_2$H$_3$O$_2$)$_2$	nickel(II) acetate
	Ni(OH)$_2$	nickel(II) hydroxide
	Ni(HCO$_3$)$_2$	nickel(II) bicarbonate
	Ni$_3$(PO$_4$)$_2$	nickel(II) phosphate
	NiSO$_3$	nickel(II) sulfite
	Ni(ClO$_4$)$_2$	nickel(II) perchlorate
	NiSO$_4$	nickel(II) sulfate
	NiO	nickel(II) oxide
	NiCl$_2$	nickel(II) chloride

Hg$_2^{2+}$:	Hg$_2$CO$_3$	mercury(I) carbonate
	Hg$_2$(BrO$_3$)$_2$	mercury(I) bromate
	Hg$_2$(C$_2$H$_3$O$_2$)$_2$	mercury(I) acetate
	Hg$_2$(OH)$_2$	mercury(I) hydroxide
	Hg$_2$(HCO$_3$)$_2$	mercury(I) bicarbonate
	(Hg$_2$)$_3$(PO$_4$)$_2$	mercury(I) phosphate
	Hg$_2$SO$_3$	mercury(I) sulfite
	Hg$_2$(ClO$_4$)$_2$	mercury(I) perchlorate
	Hg$_2$SO$_4$	mercury(I) sulfate
	Hg$_2$O	mercury(I) oxide
	Hg$_2$Cl$_2$	mercury(I) chloride

Hg^{2+}:	HgCO$_3$	mercury(II) carbonate
	Hg(BrO$_3$)$_2$	mercury(II) bromate
	Hg(C$_2$H$_3$O$_2$)$_2$	mercury(II) acetate
	Hg(OH)$_2$	mercury(II) hydroxide
	Hg(HCO$_3$)$_2$	mercury(II) bicarbonate
	Hg$_3$(PO$_4$)$_2$	mercury(II) phosphate
	HgSO$_3$	mercury(II) sulfite
	Hg(ClO$_4$)$_2$	mercury(II) perchlorate
	HgSO$_4$	mercury(II) sulfate
	HgO	mercury(II) oxide
	HgCl$_2$	mercury(II) chloride

37.

Ca(NO$_3$)$_2$	CaSO$_4$	Ca(HSO$_4$)$_2$	Ca(H$_2$PO$_4$)$_2$	CaO	CaCl$_2$
Sr(NO$_3$)$_2$	SrSO$_4$	Sr(HSO$_4$)$_2$	Sr(H$_2$PO$_4$)$_2$	SrO	SrCl$_2$
NH$_4$NO$_3$	(NH$_4$)$_2$SO$_4$	NH$_4$HSO$_4$	NH$_4$H$_2$PO$_4$	(NH$_4$)$_2$O	NH$_4$Cl
Al(NO$_3$)$_3$	Al$_2$(SO$_4$)$_3$	Al(HSO$_4$)$_3$	Al(H$_2$PO$_4$)$_3$	Al$_2$O$_3$	AlCl$_3$
Fe(NO$_3$)$_3$	Fe$_2$(SO$_4$)$_3$	Fe(HSO$_4$)$_3$	Fe(H$_2$PO$_4$)$_3$	Fe$_2$O$_3$	FeCl$_3$
Ni(NO$_3$)$_2$	NiSO$_4$	Ni(HSO$_4$)$_2$	Ni(H$_2$PO$_4$)$_2$	NiO	NiCl$_2$
AgNO$_3$	Ag$_2$SO$_4$	AgHSO$_4$	AgH$_2$PO$_4$	Ag$_2$O	AgCl

Copyright © Houghton Mifflin Company. All rights reserved.

Au(NO$_3$)$_3$	Au$_2$(SO$_4$)$_3$	Au(HSO$_4$)$_3$	Au(H$_2$PO$_4$)$_3$	Au$_2$O$_3$	AuCl$_3$
KNO$_3$	K$_2$SO$_4$	KHSO$_4$	KH$_2$PO$_4$	K$_2$O	KCl
Hg(NO$_3$)$_2$	HgSO$_4$	Hg(HSO$_4$)$_2$	Hg(H$_2$PO$_4$)$_2$	HgO	HgCl$_2$
Ba(NO$_3$)$_2$	BaSO$_4$	Ba(HSO$_4$)$_2$	Ba(H$_2$PO$_4$)$_2$	BaO	BaCl$_2$

38.
- [1] e
- [2] a
- [3] a
- [4] g
- [5] g
- [6] f
- [7] g
- [8] a
- [9] e
- [10] j

39.
 a. $Al\ (13e^-) \rightarrow Al^{3+}\ (10e^-) + 3e^-$

 b. $S\ (16e^-) + 2e^- \rightarrow S^{2-}\ (18e^-)$

 c. $Cu\ (29e^-) \rightarrow Cu^+\ (28e^-) + e^-$

 d. $F\ (9e^-) + e^- \rightarrow F^-\ (10e^-)$

 e. $Zn\ (30e^-) \rightarrow Zn^{2+}\ (28e^-) + 2e^-$

 f. $P\ (15e^-) + 3e^- \rightarrow P^{3-}\ (18e^-)$

40.
 a. none likely (element 36, Kr, is a noble gas)

 b. Ga^{3+} (element 31, Ga, is in Group 3)

 c. Te^{2-} (element 52, Te, is in Group 6)

 d. Tl^{3+} (element 81, Tl, is in Group 3)

 e. Br^- (element 35, Br, is in Group 7)

 f. Fr^+ (element 87, Fr, is in Group 1)

41.
 a. Two 1+ ions are needed to balance a 2– ion, so the formula must have two Na$^+$ ions for each S^{2-} ion: Na$_2$S.

 b. One 1+ ion exactly balances a 1– ion, so the formula should have an equal number of K$^+$ and Cl$^-$ ions: KCl.

 c. One 2+ ion exactly balances a 2– ion, so the formula must have an equal number of Ba^{2+} and O^{2-} ions: BaO.

 d. One 2+ ion exactly balances a 2– ion, so the formula must have an equal number of Mg^{2+} and Se^{2-} ions: MgSe.

 e. One 2+ ion requires two 1– ions to balance charge, so the formula must have twice as many Br$^-$ ions as Cu^{2+} ions: CuBr$_2$.

Nomenclature

    f. One 3+ ion requires three 1− ions to balance charge, so the formula must have three times as many I^- ions as Al^{3+} ions: AlI_3.

    g. Two 3+ ions give a total of 6+, whereas three 2− ions will give a total of 6−. The formula then should contain two Al^{3+} ions and three O^{2-} ions: Al_2O_3.

    h. Three 2+ ions are required to balance two 3− ions, so the formula must contain three Ca^{2+} ions for every two N^{3-} ions: Ca_3N_2.

42.
- a. beryllium oxide
- b. magnesium iodide
- c. sodium sulfide
- d. aluminum oxide
- e. hydrogen chloride (gaseous); hydrochloric acid (aqueous)
- f. lithium fluoride
- g. silver(I) sulfide; usually called silver sulfide
- h. calcium hydride

43.
- a. incorrect. Si is the element silicon, not silver.
- b. incorrect. Co is the symbol for cobalt, not copper.
- c. incorrect. Hydrogen exists as the hydride ion in this compound.
- d. correct
- e. incorrect. P is just "phosphorus" not "phosphoric".

44.
- a. Since the bromide ion must have a 1− charge, the iron ion must be in the 2+ state: the name is iron(II) bromide.
- b. Since sulfide ion always has a 2− charge, the cobalt ion must be in the 2+ state: the name is cobalt(II) sulfide.
- c. Since sulfide ion always has a 2− charge, and since there are three sulfide ions present, each cobalt ion must be in the 3+ state: the name is cobalt(III) sulfide.
- d. Since oxide ion always has a 2− charge, the tin ion must be in the 4+ state: the name is tin(IV) oxide.
- e. Since chloride ion always has a 1− charge, each mercury ion must be in the 1+ state: the name is mercury(I) chloride.
- f. Since chloride ion always has a 1− charge, the mercury ion must be in the 2+ state: the name is mercury(II) chloride.

45.
- a. xenon hexafluoride
- b. oxygen difluoride
- c. arsenic triiodide
- d. dinitrogen tetraoxide (tetroxide)

e. dichlorine monoxide
f. sulfur hexafluoride

46. a. iron(III) acetate, ferric acetate
b. bromine monofluoride
c. potassium peroxide
d. silicon tetrabromide
e. copper(II) permanganate, cupric permanganate
f. calcium chromate

47. a. lithium dihydrogen phosphate
b. copper(II) cyanide
c. lead(II) nitrate
d. sodium hydrogen phosphate
e. sodium chlorite
f. cobalt(III) sulfate

48. a. $CaCl_2$
b. Ag_2O
c. Al_2S_3
d. $BeBr_2$
e. H_2S
f. KH
g. MgI_2
h. CsF

49. a. SO_2
b. N_2O
c. XeF_4
d. P_4O_{10}
e. PCl_5
f. SF_6
g. NO_2

50. a. NaH_2PO_4
b. $LiClO_4$
c. $Cu(HCO_3)_2$

Nomenclature

    d. $KC_2H_3O_2$

    e. BaO_2

    f. Cs_2SO_3

51.   a. $AgClO_4$

    b. $Co(OH)_3$

    c.   NaClO or NaOCl

    d. $K_2Cr_2O_7$

    e. NH_4NO_2

    f. $Fe(OH)_3$

    g. NH_4HCO_3

    h. $KBrO_4$

Chapter 5 Measurements and Calculations

1. 4

2. 4,512

3. a. 0.06235
    b. 7229
    c. 0.000005001
    d. 86,210

4. a. $-5; 6.7 \times 10^{-5}$
    b. $6; 9.331442 \times 10^{6}$
    c. $-4; 1 \times 10^{-4}$
    d. $4; 1.631 \times 10^{4}$

5. a. The decimal point must be moved six places to the left, so the exponent is positive 6; $9,367,421 = 9.367421 \times 10^{6}$
    b. The decimal point must be moved three places to the left, so the exponent is positive 3; $7241 = 7.241 \times 10^{3}$
    c. The decimal point must be moved four places to the right, so the exponent is negative 4; $0.0005519 = 5.519 \times 10^{-4}$
    d. The decimal point does not have to be moved, so the exponent is zero; $5.408 = 5.408 \times 10^{0}$
    e. 6.24×10^{2} is already written in standard scientific notation.
    f. The decimal point must be moved three places to the left, and the resulting exponent of positive three must be combined with the exponent of negative two in the multiplier; $6,319 \times 10^{-2} = 6.319 \times 10^{1}$
    g. The decimal point must be moved nine places to the right, so the exponent is negative nine; $0.000000007215 = 7.215 \times 10^{-9}$
    h. The decimal point must be moved one place to the right, so the exponent is negative 1; $0.721 = 7.21 \times 10^{-1}$

6. a. The decimal point must be moved two places to the right; $4.83 \times 10^{2} = 483$
    b. The decimal point must be moved four places to the left; $7.221 \times 10^{-4} = 0.0007221$
    c. The decimal point does not have to be moved; $6.1 \times 10^{0} = 6.1$
    d. The decimal point must be moved eight places to the left; $9.11 \times 10^{-8} = 0.0000000911$

Measurements and Calculations

27

e. The decimal point must be moved six places to the right;
$4.221 \times 10^6 = 4,221,000$

f. The decimal point must be moved three places to the left;
$1.22 \times 10^{-3} = 0.00122$

g. The decimal point must be moved three places to the right;
$9.999 \times 10^3 = 9,999$

h. The decimal point must be moved five places to the left;
$1.016 \times 10^{-5} = 0.00001016$

i. The decimal point must be moved five places to the right;
$1.016 \times 10^5 = 101,600$

j. The decimal point must be moved one place to the left;
$4.11 \times 10^{-1} = 0.411$

k. The decimal point must be moved four places to the right;
$9.71 \times 10^4 = 97,100$

l. The decimal point must be moved four places to the left;
$9.71 \times 10^{-4} = 0.000971$

7. To say that scientific notation is in *standard* form means that you have a number between 1 and 10, followed by an exponential term. The numbers given in this problem are *not* between 1 and 10 as written.

   a. $142.3 \times 10^3 = (1.423 \times 10^2) \times 10^3 = 1.423 \times 10^5$

   b. $0.0007741 \times 10^{-9} = (7.741 \times 10^{-4}) \times 10^{-9} = 7.741 \times 10^{-13}$

   c. $22.7 \times 10^3 = (2.27 \times 10^1) \times 10^3 = 2.27 \times 10^4$

   d. 6.272×10^{-5} is already written in standard scientific notation.

   e. $0.0251 \times 10^4 = (2.51 \times 10^{-2}) \times 10^4 = 2.51 \times 10^2$

   f. $97,522 \times 10^{-3} = (9.7522 \times 10^4) \times 10^{-3} = 9.7522 \times 10^1$

   g. $0.0000097752 \times 10^6 = (9.7752 \times 10^{-6}) \times 10^6 = 9.7752 \times 10^0$ (9.97752)

   h. $44,252 \times 10^4 = (4.4252 \times 10^4) \times 10^4 - 4.4252 \times 10^8$

8. a. $1/0.00032 = 3.1 \times 10^3$

    b. $10^3/10^{-3} = 1 \times 10^6$

    c. $10^3/10^3 = 1$ (1×10^0); any number divided by itself is unity.

    d. $1/55,000 = 1.8 \times 10^{-5}$

    e. $(10^5)(10^4)(10^{-4})/10^{-2} = 1 \times 10^7$

    f. $43.2/(4.32 \times 10^{-5}) = \dfrac{4.32 \times 10^1}{4.32 \times 10^{-5}} = 1.00 \times 10^6$

Copyright © Houghton Mifflin Company. All rights reserved.

g. $(4.32 \times 10^{-5})/432 = \dfrac{4.32 \times 10^{-5}}{4.32 \times 10^{2}} = 1.00 \times 10^{-7}$

h. $1/(10^{5})(10^{-6}) = 1/(10^{-1}) = 1 \times 10^{1}$

9.
 a. 10^{3}
 b. 10^{-2}
 c. 10^{-3}
 d. 10^{-1}
 e. 10^{-9}
 f. 10^{-6}

10.
 a. mega-
 b. milli-
 c. nano-
 d. mega-
 e. centi-
 f. micro-

11. A mile represents, by definition, a greater distance than a kilometer. Therefore 100 mi represents a greater distance than 100 km.

12. quart

13. 5.22 cm

14. 1.62 m is approximately 5 ft, 4". The woman is slightly taller.

15. c

16. d

17. c

18. d (the other units would give very large numbers for the distance)

19. Table 5.6 indicates that a dime is 1 mm thick.

$$10 \text{ cm} \times \dfrac{10 \text{ mm}}{1 \text{ cm}} \times \dfrac{1 \text{ dime}}{1 \text{ mm}} \times \dfrac{\$1}{10 \text{ dimes}} = \$10$$

20. Table 5.6 indicates that the diameter of a quarter is 2.5 cm.

$$1 \text{m} \times \dfrac{100 \text{ cm}}{1 \text{ m}} \times \dfrac{1 \text{ quarter}}{2.5 \text{ cm}} = 40 \text{ quarters}$$

Measurements and Calculations

21. When we use a measuring scale to the *limit* of precision, we *estimate* between the smallest divisions on the scale: since this is our best estimate, the last significant digit recorded is uncertain.

22. The third figure in the length of the pin is uncertain because the measuring scale of the ruler has tenths as the smallest marked scale division. The length of the pin is given as 2.85 cm (rather than any other number) to indicate that the point of the pin appears to the observer to be half way between the smallest marked scale divisions.

23. The scale of the ruler shown is only marked to the nearest tenth of a centimeter; writing 2.850 would imply that the scale was marked to the nearest hundredth of a centimeter (and that the zero in the thousandths place had been estimated).

24.
   a. four
   b. five
   c. four
   d. one
   e. three (the decimal point makes the zeroes significant)
   f. three (because the number is written in scientific notation)
   g. six
   h. five

25.
   a. probably only two
   b. infinite (a definition)
   c. infinite (a definition)
   d. probably only 1
   e. three (the race is defined to be exactly 500. miles)

26.
   a. 1,570,000 (or better, 1.57×10^6)
   b. 2.77×10^{-3}
   c. 84,600 (or better, 8.46×10^4)
   d. 0.00117
   e. 0.0776

27.
   a. 3.42×10^{-4}
   b. 1.034×10^4
   c. 1.7992×10^1
   d. 3.37×10^5

28. two significant figures (based on 0.0043 having two significant figures)

29. three

30. only one (based on 121.2 being known only to the first decimal place)

31. none

32. a. 641.0 (the answer can only be given to one decimal place, since 212.7 and 26.7 are only given to one decimal place)

   b. 1.327 (the answer can only be given to three decimal places, since 0.221 is only given to three decimal places)

   c. 77.34 (the answer can only be given to two decimal places, since 26.01 is only given to two decimal places)

   d. Before performing the calculation, the numbers have to be converted so that they contain the same power of ten.
   $2.01 \times 10^2 + 3.014 \times 10^3 = 2.01 \times 10^2 + 30.14 \times 10^2 = 32.15 \times 10^2$
   This answer should then be converted to *standard* scientific notation,
   $32.15 \times 10^2 = 3.215 \times 10^3 = 3,215$.

33. a. 124 (the answer can only be given to three significant figures because 0.995 is only given to three significant figures)

   b. 1.995×10^{-23} (the answer can only be given to four significant figures because 6.022×10^{23} is only given to four significant figures)

   c. 1.14×10^{-2} (the answer can only be given to three significant figures because 0.500 is only given to three significant figures)

   d. 5.3×10^{-4} (the answer can only be given to three significant figures because 0.15 is only given to two significant figures)

34. a. $(2.0944 + 0.0003233 + 12.22)/7.001 = (14.3147233)/7.001 = 2.045$

   b. $(1.42 \times 10^2 + 1.021 \times 10^3)/(3.1 \times 10^{-1}) = (142 + 1021)/(3.1 \times 10^{-1}) = (1163)/(3.1 \times 10^{-1}) = 3751 = 3.8 \times 10^3$

   c. $(9.762 \times 10^{-3})/(1.43 \times 10^2 + 4.51 \times 10^1) = (9.762 \times 10^{-3})/(143 + 45.1) = (9.762 \times 10^{-3})/(188.1) = 5.19 \times 10^{-5}$

   d. $(6.1982 \times 10^{-4})^2 = (6.1982 \times 10^{-4})(6.1982 \times 10^{-4}) = 3.8418 \times 10^{-7}$

35. an infinite number (a definition)

36. $\dfrac{1 \text{ mi}}{1760 \text{ yd}}$; $\dfrac{1760 \text{ yd}}{1 \text{ mi}}$

37. $\dfrac{\$0.79}{1 \text{ lb}}$

38. $\dfrac{1 \text{ lb}}{\$0.79}$

Measurements and Calculations

39.
a. $2.23 \text{ m} \times \dfrac{1.094 \text{ yd}}{1 \text{ m}} = 2.44 \text{ yd}$

b. $46.2 \text{ yd} \times \dfrac{1 \text{ m}}{1.094 \text{ yd}} = 42.2 \text{ m}$

c. $292 \text{ cm} \times \dfrac{1 \text{ in.}}{2.54 \text{ cm}} = 115 \text{ in.}$

d. $881.2 \text{ in.} \times \dfrac{2.54 \text{ cm}}{1 \text{ in.}} = 2238 \text{ cm}$

e. $1043 \text{ km} \times \dfrac{1 \text{ mi}}{1.6093 \text{ km}} = 648.1 \text{ mi}$

f. $445.5 \text{ mi} \times \dfrac{1.6093 \text{ km}}{1 \text{ mi}} = 716.9 \text{ km}$

g. $36.2 \text{ m} \times \dfrac{1 \text{ km}}{1000 \text{ m}} = 0.0362 \text{ km}$

h. $0.501 \text{ km} \times \dfrac{1000 \text{ m}}{1 \text{ km}} \times \dfrac{100 \text{ cm}}{1 \text{ m}} = 5.01 \times 10^4 \text{ cm}$

40.
a. $254.3 \text{ g} \times \dfrac{1 \text{ kg}}{1000 \text{ g}} = 0.2543 \text{ kg}$

b. $2.75 \text{ kg} \times \dfrac{1000 \text{ g}}{1 \text{ kg}} = 2.75 \times 10^3 \text{ g}$

c. $2.75 \text{ kg} \times \dfrac{1 \text{ lb}}{0.45359 \text{ kg}} = 6.06 \text{ lb}$

d. $2.75 \text{ kg} \times \dfrac{1 \text{ lb}}{0.45359 \text{ kg}} \times \dfrac{16 \text{ oz}}{1 \text{ lb}} = 97.0 \text{ oz}$

e. $534.1 \text{ g} \times \dfrac{1 \text{ lb}}{453.59 \text{ g}} = 1.177 \text{ lb}$

f. $1.75 \text{ lb} \times \dfrac{453.59 \text{ g}}{1 \text{ lb}} = 794 \text{ g}$

g. $8.7 \text{ oz} \times \dfrac{1 \text{ lb}}{16 \text{ oz}} \times \dfrac{453.59 \text{ g}}{1 \text{ lb}} = 2.5 \times 10^2 \text{ g}$

h. $45.9 \text{ g} \times \dfrac{1 \text{ lb}}{453.59 \text{ g}} \times \dfrac{16 \text{ oz}}{1 \text{ lb}} = 1.62 \text{ oz}$

41. $\$20.00 \times \dfrac{\text{DM } 1.74}{\$1.00} = \text{DM } 34.80$ (assuming the exchange rate is exact)

$\text{DM } 100.0 \times \dfrac{\$1.00}{\text{DM } 1.74} = \$57.47$ (assuming the exchange rate is exact)

Copyright © Houghton Mifflin Company. All rights reserved.

42. 190 mi = 1.9×10^2 mi to two significant figures

$$1.9 \times 10^2 \text{ mi} \times \frac{1 \text{ km}}{0.62137 \text{ mi}} = 3.1 \times 10^2 \text{ km}$$

$$3.1 \times 10^2 \text{ km} \times \frac{1000 \text{ m}}{1 \text{ km}} = 3.1 \times 10^5 \text{ m}$$

$$1.9 \times 10^2 \text{ mi} \times \frac{5,280 \text{ ft}}{1 \text{ mi}} = 1.0 \times 10^6 \text{ ft}$$

43. To decide which train is faster, both speeds must be expressed in the *same unit* of distance (either miles or kilometers)

$$\frac{225 \text{ km}}{1 \text{ hr}} \times \frac{1 \text{ mi}}{1.6093 \text{ km}} = 140. \text{ mi/hr}$$

So the Boston-New York trains will be faster.

44. 212°F; 100°C

45. 100

46. Fahrenheit (F)

47. $t_K = t_C + 273$ $\quad\quad$ $t_C = (t_F - 32)/1.80$

    a. $-155 + 273 = 118$ K

    b. $200 + 273 = 473$ K

    c. $-52 + 273 = 221$ K

    d. 101°F = 38.3°C; 38.3 + 273 = 311 K

    e. -52°F = -46.6°C; $-46.6 + 273 = 226$ K

    f. $-196 + 273 = 77$ K

48. $t_C = t_K - 273$

    a. $275 - 273 = 2$°C

    b. $445 - 273 = 172$°C

    c. $0 - 273 = -273$°C

    d. $77 - 273 = -196$°C

    e. $10,000. - 273 = 9727$°C

    f. $2 - 273 = -271$°C

49. a. $1.80(-40) + 32 = -40$°F

    b. $(-40 - 32)/180 = -40$°C

Measurements and Calculations

      c.   232 − 273 = −41°C

      d.   232 K = −41°C; 1.80(−41) + 32 = −42°F

50.   a. $t_C = (t_F - 32)/1.80 = (-201°F - 32)/1.80 = (-233)/1.80 = -129.4°C$

               −129.4°C + 273 = 143.6 = 144 K

      b.   −201°C + 273 = 72 K

      c. $t_F = 1.80(t_C) + 32 = 1.80(351°C) + 32 = 664°F$

      d. $t_C = (t_F - 32)/1.80 = (-150°F - 32)/1.80 = -101°C$

51.   volume

52.   lead

53.   low

54.   Density is a *characteristic* property of a pure substance; all samples of the same pure substance have the *same* density.

55.   aluminum (2.70 g/cm³)

56. $\text{density} = \dfrac{\text{mass}}{\text{volume}}$

      a.   m = 4.53 kg = 4530 g

               $d = \dfrac{4530 \text{ g}}{225 \text{ cm}^3} = 20.1 \text{ g/cm}^3$

      b.   v = 25.0 mL = 25.0 cm³

               $d = \dfrac{26.3 \text{ g}}{25.0 \text{ cm}^3} = 1.05 \text{ g/cm}^3$

      c.   m = 1.00 lb = 453.59 g

               $d = \dfrac{453.59 \text{ g}}{500. \text{ cm}^3} = 0.907 \text{ g/cm}^3$

      d.   m = 352 mg = 0.352 g

               $d = \dfrac{0.352 \text{ g}}{0.271 \text{ cm}^3} = 1.30 \text{ g/cm}^3$

57. $d = \dfrac{75.2 \text{ g}}{89.2 \text{ mL}} = 0.843 \text{ g/mL}$

58. $m = 1.45 \text{ kg} \times \dfrac{10^3 \text{ g}}{1 \text{ kg}} = 1.45 \times 10^3$

    $d = \dfrac{1.45 \times 10^3 \text{ g}}{542 \text{ mL}} = 2.68 \text{ g/mL}$

59. $m = 3.5 \text{ lb} \times \dfrac{453.59 \text{ g}}{1 \text{ lb}} = 1.59 \times 10^3 \text{ g}$

    $v = 1.2 \times 10^4 \text{ in}^3 \times \left(\dfrac{2.54 \text{ cm}}{1 \text{ in}}\right)^3 = 1.97 \times 10^5 \text{ cm}^3$

    $d = \dfrac{1.59 \times 10^3 \text{ g}}{1.97 \times 10^5 \text{ cm}^3} = 8.1 \times 10^{-3} \text{ g/cm}^3$

    The material will float.

60. $5.25 \text{ g} \times \dfrac{1 \text{ cm}^3}{10.5 \text{ g}} = 0.500 \text{ cm}^3 = 0.500 \text{ mL}$

    $11.2 \text{ mL} + 0.500 \text{ mL} = 11.7 \text{ mL}$

61. a. $50.0 \text{ g} \times \dfrac{1 \text{ cm}^3}{2.16 \text{ g}} = 23.1 \text{ cm}^3$

    b. $50.0 \text{ g} \times \dfrac{1 \text{ cm}^3}{13.6 \text{ g}} = 3.68 \text{ cm}^3$

    c. $50.0 \text{ g} \times \dfrac{1 \text{ cm}^3}{0.880 \text{ g}} = 56.8 \text{ cm}^3$

    d. $50.0 \text{ g} \times \dfrac{1 \text{ cm}^3}{10.5 \text{ g}} = 4.76 \text{ cm}^3$

62. a. $50.0 \text{ cm}^3 \times \dfrac{19.32 \text{ g}}{1 \text{ cm}^3} = 966 \text{ g}$

    b. $50.0 \text{ cm}^3 \times \dfrac{7.87 \text{ g}}{1 \text{ cm}^3} = 394 \text{ g}$

    c. $50.0 \text{ cm}^3 \times \dfrac{11.34 \text{ g}}{1 \text{ cm}^3} = 567 \text{ g}$

    d. $50.0 \text{ cm}^3 \times \dfrac{2.70 \text{ g}}{1 \text{ cm}^3} = 135 \text{ g}$

63. a. centimeters

    b. meters

    c. kilometers

d. centimeters

e. millimeters

64. $45 \text{ mi} \times \dfrac{1.6093}{1 \text{ mi}} = 72.4 \text{ km}$

$38 \text{ mi} \times \dfrac{1.6093}{1 \text{ mi}} = 61.2 \text{ km}$

1 gal = 3.7854 L

highway: 72.4 km/3.7854 L = 19 km/L

city: 61.2 km/3.7854 L = 16 km/L

65. $1 \text{ lb} \times \dfrac{1 \text{ kg}}{2.2 \text{ lb}} \times \dfrac{\$1}{F5} \times \dfrac{11.5F}{1 \text{ kg}} = \$1$

66. $15.6 \text{ g} \times \dfrac{1 \text{ capsule}}{0.65 \text{ g}} = 24 \text{ capsules}$

67. $v = 4/3(\pi r^3) = 4/3(3.1416)(0.5 \text{ cm})^3 = 0.52 \text{ cm}^3$

$d = \dfrac{2.0 \text{ g}}{0.52 \text{ cm}^3} = 3.8 \text{ g/cm}^3$ (the ball will sink)

68. for ethanol, $100. \text{ mL} \times \dfrac{0.785 \text{ g}}{1 \text{ mL}} = 78.5 \text{ g}$

for benzene, $1000 \text{ mL} \times \dfrac{0.880 \text{ g}}{1 \text{ mL}} = 880. \text{ g}$

total mass, 78.5 + 880. = 959 g

69. a. 4; positive

b. 6; negative

c. 0; zero

d. 5; positive

e. 2; negative

70. 4.25 g (425 mg = 0.425 g)

71. 2.8 (the hundredths place is estimated)

72. a. one

b. one

c. four

d. two

e. infinite (definition)

f. one

73. $50.0 \text{ g} \times \dfrac{1 \text{ mL}}{1.31 \text{ g}} = 38.2 \text{ g}$

74. Volume = 21.6 mL − 12.7 mL = 8.9 mL

$d = \dfrac{33.42 \text{ g}}{8.9 \text{ mL}} = 3.8 \text{ g/mL}$

Chapter 6 Chemical Composition

1. 100 washers × $\dfrac{0.110 \text{ g}}{1 \text{ washer}}$ = 11.0 g (assuming 100 washers is exact).

   100. g × $\dfrac{1 \text{ washer}}{0.110 \text{ g}}$ = 909 washers

2. 500. g × $\dfrac{1 \text{ cork}}{1.63 \text{ g}}$ = 306.7 = 307 corks

   500. g × $\dfrac{1 \text{ stopper}}{4.31 \text{ g}}$ = 116 stoppers

   1 kg (1000 g) of corks contains (1000 g × $\dfrac{1 \text{ cork}}{1.63 \text{ g}}$) = 613.49 = 613 corks

   613 stoppers would weigh (613 stoppers × $\dfrac{4.31 \text{ g}}{1 \text{ stopper}}$) = 2644 g = 2640 g

   The ratio of the mass of a stopper to the mass of a cork is (4.31 g/1.63 g). So the mass of stoppers that contains the same number of stoppers as there are corks in 1000 g of corks is

   1000g × $\dfrac{4.31 \text{ g}}{1.63 \text{ g}}$ = 2644 g = 2640 g

3. The *atomic mass unit* (amu) is a unit of mass defined by scientists to more simply describe relative masses on an atomic or molecular scale. One amu is equivalent to 1.66×10^{-24} g.

4. We use the average atomic mass of an element when performing calculations because the average mass takes into account the individual masses and relative abundance of all the isotopes of an element.

5.
   a. 635 H atoms × $\dfrac{1.008 \text{ amu}}{1 \text{ H atom}}$ = 640. amu

   b. 1.261×10^4 W atoms × $\dfrac{183.9 \text{ amu}}{1 \text{ W atom}}$ = 2.319×10^6 amu

   c. 42 K atoms × $\dfrac{39.10 \text{ amu}}{1 \text{ K atom}}$ = 1642 amu

   d. 7.213×10^{23} N atoms × $\dfrac{14.01 \text{ amu}}{1 \text{ N atom}}$ = 1.011×10^{25} amu

   e. 891 Fe atoms × $\dfrac{55.85 \text{ amu}}{1 \text{ Fe atom}}$ = 4.976×10^4 amu

6. a. $10.81 \text{ amu} \times \dfrac{1 \text{ B atom}}{10.81 \text{ amu}} = 1.000 \text{ atom} = 1 \text{ B atom}$

   b. $320.7 \text{ amu} \times \dfrac{1 \text{ S atom}}{32.07 \text{ amu}} = 10 \text{ S atoms}$

   c. $19{,}691 \text{ amu} \times \dfrac{1 \text{ Au atom}}{197.97 \text{ amu}} = 100.00 \text{ Au atoms} = 100 \text{ Au atoms}$

   d. $19{,}695 \text{ amu} \times \dfrac{1 \text{ Xe atom}}{131.3 \text{ amu}} = 150.0 \text{ Xe atoms} = 150 \text{ Xe atoms}$

   e. $3588.3 \text{ amu} \times \dfrac{1 \text{ Al atom}}{26.98 \text{ amu}} = 133.0 \text{ Al atoms} = 133 \text{ Al atoms}$

7. $8274 \text{ amu} \times \dfrac{1 \text{ S atom}}{32.07 \text{ amu}} = 258 \text{ S atoms}$

   $5.213 \times 10^{24} \text{ S atoms} \times \dfrac{32.07 \text{ amu}}{1 \text{ S atom}} = 1.672 \times 10^{26} \text{ amu}$

8. 1.204×10^{24}

9. Avogadro's number (6.022×10^{23})

10. The ratio of the atomic mass of Ca to the atomic mass of Mg is (40.08 amu/24.31 amu), and the masses of calcium are given by

    $12.16 \text{ g Mg} \times \dfrac{40.08 \text{ amu}}{24.31 \text{ amu}} = 20.05 \text{ g Ca}$

    $24.31 \text{ g Mg} \times \dfrac{40.08 \text{ amu}}{24.31 \text{ amu}} = 40.08 \text{ g Ca}$

11. The ratio of the atomic mass of Co to the atomic mass of F is (58.93 amu/19.00 amu), and the mass of cobalt is given by

    $57.0 \text{ g} \times \dfrac{58.93 \text{ amu}}{19.00 \text{ amu}} = 177 \text{ g Co}$

12. $1 \text{ mol O} = 16.00 \text{ g O} = 6.02 \times 10^{23} \text{ O atoms}$

    $1 \text{ O atom} \times \dfrac{16.00 \text{ g O}}{6.022 \times 10^{23} \text{ O atoms}} = 2.66 \times 10^{-23} \text{ g}$

13. $0.50 \text{ mol O atoms} \times \dfrac{16.00 \text{ g O}}{1 \text{ mol}} = 8.0 \text{ g O}$

    $4 \text{ mol H atoms} \times \dfrac{1.008 \text{ g}}{1 \text{ mol}} = 4 \text{ g H}$

    Half a mole of O atoms weighs more than 4 moles of H atoms.

Chemical Composition

14.
a. $26.2 \text{ g Au} \times \dfrac{1 \text{ mol Au}}{197.0 \text{ g}} = 0.133 \text{ mol Au}$

b. $41.5 \text{ g Ca} \times \dfrac{1 \text{ mol Ca}}{40.08 \text{ g}} = 1.04 \text{ mol Ca}$

c. $335 \text{ mg Ba} \times \dfrac{1 \text{ g}}{10^3 \text{ mg}} \times \dfrac{1 \text{ mol Ba}}{137.3 \text{ g}} = 2.44 \times 10^{-3} \text{ mol Ba}$

d. $1.42 \times 10^{-3} \text{ g Pd} \times \dfrac{1 \text{ mol Pd}}{106.4 \text{ g}} = 1.33 \times 10^{-5} \text{ mol Pd}$

e. $3.05 \times 10^{-5} \text{ µg Ni} \times \dfrac{1 \text{ g}}{10^6 \text{ µg}} \times \dfrac{1 \text{ mol Ni}}{58.70 \text{ g}} = 5.20 \times 10^{-13} \text{ mol Ni}$

f. $1.00 \text{ lb Fe} \times \dfrac{453.59 \text{ g}}{1 \text{ lb}} \times \dfrac{1 \text{ mol Fe}}{55.85 \text{ g}} = 8.12 \text{ mol Fe}$

g. $12.01 \text{ g C} \times \dfrac{1 \text{ mol C}}{12.01 \text{ g}} = 1.000 \text{ mol C}$

15.
a. $2.00 \text{ mol Fe} \times \dfrac{55.85 \text{ g}}{1 \text{ mol}} = 112 \text{ g Fe}$

b. $0.521 \text{ mol Ni} \times \dfrac{58.70 \text{ g}}{1 \text{ mol}} = 30.6 \text{ g Ni}$

c. $1.23 \times 10^{-3} \text{ mol Pt} \times \dfrac{195.1 \text{ g}}{1 \text{ mol}} = 0.240 \text{ g Pt}$

d. $72.5 \text{ mol Pb} \times \dfrac{207.2 \text{ g}}{1 \text{ mol}} = 1.50 \times 10^4 \text{ g Pb}$

e. $0.00102 \text{ mol Mg} \times \dfrac{24.31 \text{ g}}{1 \text{ mol}} = 0.0248 \text{ g Mg}$

f. $4.87 \times 10^3 \text{ mol Al} \times \dfrac{26.98 \text{ g}}{1 \text{ mol}} = 1.31 \times 10^5 \text{ g Al}$

g. $211.5 \text{ mol Li} \times \dfrac{6.941 \text{ g}}{1 \text{ mol}} = 1468 \text{ g Li}$

h. $1.72 \times 10^{-6} \text{ mol Na} \times \dfrac{22.99 \text{ g}}{1 \text{ mol}} = 3.95 \times 10^{-5} \text{ g Na}$

16.
a. $0.00103 \text{ g Co} \times \dfrac{6.022 \times 10^{23} \text{ Co atoms}}{58.93 \text{ g Co}} = 1.05 \times 10^{19} \text{ Co atoms}$

b. $0.00103 \text{ mol Co} \times \dfrac{6.022 \times 10^{23} \text{ Co atoms}}{1 \text{ mol}} = 6.20 \times 10^{20} \text{ Co atoms}$

Copyright © Houghton Mifflin Company. All rights reserved.

c.   2.75 g cobalt × $\dfrac{1 \text{ mol}}{58.93 \text{ g Co}}$ = 0.0467 mol Co

d. 5.99×10^{21} Co atoms × $\dfrac{1 \text{ mol}}{6.022 \times 10^{23} \text{ Co atoms}}$ = 0.00995 mol Co

e.   4.23 mol Co × $\dfrac{58.93 \text{ g Co}}{1 \text{ mol Co}}$ = 249 g Co

f.   4.23 mol Co × $\dfrac{6.022 \times 10^{23} \text{ Co atoms}}{1 \text{ mol Co}}$ = 2.55×10^{24} Co atoms

g.   4.23 g Co × $\dfrac{6.022 \times 10^{23} \text{ Co atoms}}{58.93 \text{ g Co}}$ = 4.32×10^{22} Co atoms

17. molar mass

18. adding together (summing)

19.
a.
mass of 3 mol Na =	3 (22.99 g) =	68.97 g
mass of 1 mol N =	1 (14.01 g) =	14.01 g
molar mass of Na_3N =		82.98 g

b.
mass of 1 mol C =	12.01 g =	12.01 g
mass of 2 mol S =	2 (32.07 g) =	64.14 g
molar mass of CS_2 =		76.15 g

c.
mass of 1 mol N =	14.01 g =	14.01 g
mass of 4 mol H =	4 (1.008 g) =	4.032 g
mass of 1 mol Br =	79.90 g =	79.90 g
molar mass of NH_4Br =		97.942 g = 97.94 g

d.
mass of 2 mol C =	2 (12.01 g) =	24.02 g
mass of 6 mol H =	6 (1.008 g) =	6.048 g
mass of 1 mol O =	16.00 g =	16.00 g
molar mass of C_2H_6O =		46.068 g = 46.07 g

e.
mass of 2 mol H =	2 (1.008 g) =	2.016 g
mass of 1 mol S =	32.07 g =	32.07 g
mass of 3 mol O =	3 (16.00 g) =	48.00 g
molar mass of H_2SO_3 =		82.086 g = 82.09 g

f.
mass of 2 mol H =	2 (1.008 g) =	2.016 g
mass of 1 mol S =	32.07 g =	32.07 g
mass of 4 mol O =	4 (16.00 g) =	64.00 g
molar mass of H_2SO_4 =		98.086 g = 98.09 g

20.
a.
mass of 1 mol Ba =	137.3 g =	137.3 g
mass of 2 mol Cl =	2 (35.45 g) =	70.90 g
mass of 8 mol O =	8 (16.00 g) =	128.0 g
molar mass of $Ba(ClO_4)_2$ =		336.2 g

Chemical Composition

b.
mass of 1 mol Mg =   24.31 g =   24.31 g
mass of 1 mol S =   32.07 g =   32.07 g
mass of 4 mol O =   4 (16.00 g) =   64.00 g
molar mass of $MgSO_4$ =   120.38 g

c.
mass of 1 mol Pb =   207.2 g =   207.2 g
mass of 2 mol Cl =   2 (35.45 g) =   79.90 g
molar mass of $PbCl_2$ =   278.1 g

d.
mass of 1 mol Cu =   63.55 g =   63.55 g
mass of 2 mol N =   2 (14.01 g) =   28.02 g
mass of 6 mol O =   6 (16.00 g) =   96.00 g
molar mass of $Cu(NO_3)_2$ =   187.57 g

e.
mass of 1 mol Sn =   118.7 g =   118.7 g
mass of 4 mol Cl =   4 (35.45 g) =   141.80 g
molar mass of $SnCl_4$ =   260.5 g

f.
mass of 6 mol C =   6 (12.01 g) =   72.06 g
mass of 6 mol H =   6 (1.008 g) =   6.048 g
mass of 1 mol O =   16.00 g =   16.00 g
molar mass of C_6H_6O =   94.11 g

21.
a. molar mass of SO_3 = 80.07 g

$$49.2 \text{ mg } SO_3 \times \frac{1 \text{ g}}{1000 \text{ mg}} \times \frac{1 \text{ mol}}{80.07 \text{ g}} = 6.14 \times 10^{-4} \text{ mol } SO_3$$

b. molar mass of PbO_2 = 239.2 g

$$7.44 \times 10^4 \text{ kg } PbO_2 \times \frac{1000 \text{ g}}{1 \text{ kg}} \times \frac{1 \text{ mol}}{239.2 \text{ g}} = 3.11 \times 10^5 \text{ mol } PbO_2$$

c. molar mass of $CHCl_3$ = 119.37 g

$$59.1 \text{ g } CHCl_3 \times \frac{1 \text{ mol}}{119.37 \text{ g}} = 0.495 \text{ mol } CHCl_3$$

d. molar mass of $C_2H_3Cl_3$ = 133.39 g

$$3.27 \text{ μg} \times \frac{1 \text{ g}}{10^6 \text{ μg}} \times \frac{1 \text{ mol}}{133.39 \text{ g}} = 2.45 \times 10^{-8} \text{ mol } C_2H_3Cl_3$$

e. molar mass of LiOH = 23.95 g

$$4.01 \text{ g LiOH} \times \frac{1 \text{ mol}}{23.95 \text{ g}} = 0.167 \text{ mol LiOH}$$

22. a. molar mass of NaH_2PO_4 = 120.0 g

$$4.26 \times 10^{-3} \text{ g } NaH_2PO_4 \times \frac{1 \text{ mol}}{120.0 \text{ g}} = 3.55 \times 10^{-5} \text{ mol } NaH_2PO_4$$

b. molar mass of $CuCl$ = 99.00 g

$$521 \text{ g } CuCl \times \frac{1 \text{ mol}}{99.00 \text{ g}} = 5.26 \text{ mol } CuCl$$

c. molar mass of Fe = 55.85 g

$$151 \text{ kg } Fe \times \frac{1000 \text{ g}}{1 \text{ kg}} \times \frac{1 \text{ mol}}{55.85 \text{ g}} = 2.70 \times 10^3 \text{ mol } Fe$$

d. molar mass of SrF_2 = 125.6 g

$$8.76 \text{ g } SrF_2 \times \frac{1 \text{ mol}}{125.6 \text{ g}} = 0.0697 \text{ mol } SrF_2$$

e. molar mass of Al = 26.98 g

$$1.26 \times 10^4 \text{ g } Al \times \frac{1 \text{ mol}}{26.98 \text{ g}} = 467 \text{ mol } Al$$

23. a. molar mass of AlI_3 = 407.7 g

$$1.50 \text{ mol } AlI_3 \times \frac{407.7 \text{ g}}{1 \text{ mol}} = 612 \text{ g of } AlI_3$$

b. molar mass of C_6H_6 = 78.11 g

$$1.91 \times 10^{-3} \text{ mol } C_6H_6 \times \frac{78.11 \text{ g}}{1 \text{ mol}} = 0.149 \text{ g } C_6H_6$$

c. molar mass of $C_6H_{12}O_6$ = 180.2 g

$$4.00 \text{ mol } C_6H_{12}O_6 \times \frac{180.2 \text{ g}}{1 \text{ mol}} = 721 \text{ g } C_6H_{12}O_6$$

d. molar mass of C_2H_5OH = 46.07 g

$$4.56 \times 10^5 \text{ mol } C_2H_5OH \times \frac{46.07 \text{ g}}{1 \text{ mol}} = 2.10 \times 10^7 \text{ g } C_2H_5OH$$

e. molar mass of $Ca(NO_3)_2$ = 164.1 g

$$2.27 \text{ mol } Ca(NO_3)_2 \times \frac{164.1 \text{ g}}{1 \text{ mol}} = 373 \text{ g } Ca(NO_3)_2$$

Chemical Composition

24. a. molar mass of CO_2 = 44.01 g

$$1.27 \text{ mmol} \times \frac{1 \text{ mol}}{10^3 \text{ mmol}} \times \frac{44.01 \text{ g}}{1 \text{ mol}} = 0.0559 \text{ g } CO_2$$

b. molar mass of NCl_3 = 120.4 g

$$4.12 \times 10^3 \text{ mol } NCl_3 \times \frac{120.4 \text{ g}}{1 \text{ mol}} = 4.96 \times 10^5 \text{ g } NCl_3$$

c. molar mass of NH_4NO_3 = 80.05 g

$$0.00451 \text{ mol } NH_4NO_3 \times \frac{80.05 \text{ g}}{1 \text{ mol}} = 0.361 \text{ g } NH_4NO_3$$

d. molar mass of H_2O = 18.02 g

$$18.0 \text{ mol } H_2O \times \frac{18.02 \text{ g}}{1 \text{ mol}} = 324 \text{ g } H_2O$$

e. molar mass of $CuSO_4$ = 159.6 g

$$62.7 \text{ mol } CuSO_4 \times \frac{159.6 \text{ g}}{1 \text{ mol}} = 1.00 \times 10^4 \text{ g } CuSO_4$$

25. a. $6.37 \text{ mol } CO \times \frac{6.022 \times 10^{23} \text{ molecules}}{1 \text{ mol}} = 3.84 \times 10^{24} \text{ molecules } CO$

b. molar mass of CO = 28.01 g

$$6.37 \text{ g} \times \frac{1 \text{ mol}}{28.01 \text{ g}} \times \frac{6.022 \times 10^{23} \text{ molecules}}{1 \text{ mol}} = 1.37 \times 10^{23} \text{ molecules } CO$$

c. molar mass of H_2O = 18.02 g

$$2.62 \times 10^{-6} \text{ g} \times \frac{6.022 \times 10^{23} \text{ molecules}}{18.02 \text{ g}} = 8.76 \times 10^{16} \text{ molecules } H_2O$$

d. $2.62 \times 10^{-6} \text{ mol} \times \frac{6.022 \times 10^{23} \text{ molecules}}{1 \text{ mol}} = 1.58 \times 10^{18} \text{ molecules } H_2O$

e. molar mass of C_6H_6 = 78.11 g

$$5.23 \text{ g} \times \frac{6.022 \times 10^{23} \text{ molecules}}{78.11 \text{ g}} = 4.03 \times 10^{22} \text{ molecules } C_6H_6$$

26. a. molar mass of Na_2SO_4 = 142.1 g

$$2.01 \text{ g } Na_2SO_4 \times \frac{1 \text{ mol } Na_2SO_4}{142.1 \text{ g}} \times \frac{1 \text{ mol S}}{1 \text{ mol } Na_2SO_4} = 0.0141 \text{ mol S}$$

b. molar mass of Na_2SO_3 = 126.1 g

$$2.01 \text{ g Na}_2\text{SO}_3 \times \frac{1 \text{ mol Na}_2\text{SO}_3}{126.1 \text{ g}} \times \frac{1 \text{ mol S}}{1 \text{ mol Na}_2\text{SO}_3} = 0.0159 \text{ mol S}$$

c. molar mass of Na_2S = 78.05 g

$$2.01 \text{ g Na}_2\text{S} \times \frac{1 \text{ mol Na}_2\text{S}}{78.05 \text{ g}} \times \frac{1 \text{ mol S}}{1 \text{ mol Na}_2\text{S}} = 0.0258 \text{ mol S}$$

d. molar mass of $Na_2S_2O_3$ = 158.1 g

$$2.01 \text{ g Na}_2\text{S}_2\text{O}_3 \times \frac{1 \text{ mol Na}_2\text{S}_2\text{O}_3}{158.1 \text{ g}} \times \frac{2 \text{ mol S}}{1 \text{ mol Na}_2\text{S}_2\text{O}_3} = 0.0254 \text{ mol S}$$

27. a.
mass of Na present = 2 (22.99 g) = 45.98 g
mass of S present = 32.07 g = 32.07 g
mass of O present = 4 (16.00 g) = 64.00 g
molar mass of Na_2SO_4 = 142.05 g

$$\% \text{ Na} = \frac{45.98 \text{ g Na}}{142.05 \text{ g}} \times 100 = 32.37\% \text{ Na}$$

$$\% \text{ S} = \frac{32.07 \text{ g S}}{142.05 \text{ g}} \times 100 = 22.58\% \text{ S}$$

$$\% \text{ O} = \frac{64.00 \text{ g O}}{142.05 \text{ g}} \times 100 = 45.05\% \text{ O}$$

b.
mass of Na present = 2 (22.99 g) = 45.98 g
mass of S present = 32.07 g = 32.07 g
mass of O present = 3 (16.00 g) = 48.00 g
molar mass of Na_2SO_3 = 126.05 g

$$\% \text{ Na} = \frac{45.98 \text{ g Na}}{126.05 \text{ g}} \times 100 = 36.48\% \text{ Na}$$

$$\% \text{ S} = \frac{32.07 \text{ g S}}{126.05 \text{ g}} \times 100 = 25.44\% \text{ S}$$

$$\% \text{ O} = \frac{48.00 \text{ g O}}{126.05 \text{ g}} \times 100 = 38.08\% \text{ O}$$

c.
mass of Na present = 2 (22.99 g) = 45.98 g
mass of S present = 32.07 g = 32.07 g
molar mass of Na_2S = 78.05 g

$$\% \text{ Na} = \frac{45.98 \text{ g Na}}{78.05 \text{ g}} \times 100 = 58.91\% \text{ Na}$$

$$\% \text{ S} = \frac{32.07 \text{ g S}}{78.05 \text{ g}} \times 100 = 41.09\% \text{ S}$$

Chemical Composition

d. mass of Na present =   2 (22.99 g) =   45.98 g
    mass of S present =   2 (32.07 g) =   64.14 g
    mass of O present =   3 (16.00 g) =   48.00 g
    molar mass of $Na_2S_2O_3$ =   158.12 g

$$\% \text{ Na} = \frac{45.98 \text{ g Na}}{158.12 \text{ g}} \times 100 = 29.08\% \text{ Na}$$

$$\% \text{ S} = \frac{64.14 \text{ g S}}{158.12 \text{ g}} \times 100 = 40.56\% \text{ S}$$

$$\% \text{ O} = \frac{48.00 \text{ g O}}{158.12 \text{ g}} \times 100 = 30.36\% \text{ O}$$

e. mass of K present =   3 (39.10 g) =   117.3 g
    mass of P present =   30.97 g =   30.97 g
    mass of O present =   4 (16.00 g) =   64.00 g
    molar mass of K_3PO_4 =   212.3 g

$$\% \text{ K} = \frac{117.3 \text{ g K}}{212.3 \text{ g}} \times 100 = 55.25\% \text{ K}$$

$$\% \text{ P} = \frac{30.97 \text{ g P}}{212.3 \text{ g}} \times 100 = 14.59\% \text{ P}$$

$$\% \text{ O} = \frac{64.00 \text{ g O}}{212.3 \text{ g}} \times 100 = 30.15\% \text{ O}$$

f. mass of K present =   2 (39.10 g) =   78.20 g
    mass of H present =   1.008 g =   1.008 g
    mass of P present =   30.97 g =   30.97 g
    mass of O present =   4 (16.00 g) =   64.00 g
    molar mass of K_2HPO_4 =   174.178 g = 174.18 g

$$\% \text{ K} = \frac{78.20 \text{ g K}}{174.18 \text{ g}} \times 100 = 44.90\% \text{ K}$$

$$\% \text{ H} = \frac{1.008 \text{ g H}}{174.18 \text{ g}} \times 100 = 0.5787\% \text{ H}$$

$$\% \text{ P} = \frac{30.97 \text{ g P}}{174.18 \text{ g}} \times 100 = 17.78\% \text{ P}$$

$$\% \text{ O} = \frac{64.00 \text{ g O}}{174.18 \text{ g}} \times 100 = 36.74\% \text{ O}$$

g. mass of K present =   39.10 g =   39.10 g
    mass of H present =   2 (1.008 g) =   2.016 g
    mass of P present =   30.97 g =   30.97 g
    mass of O present =   4 (16.00 g) =   64.00 g
    molar mass of KH_2PO_4 =   136.09 g

Copyright © Houghton Mifflin Company. All rights reserved.

$\% K = \dfrac{39.10 \text{ g K}}{136.09 \text{ g}} \times 100 = 28.73\% \text{ K}$

$\% H = \dfrac{2.016 \text{ g H}}{136.09 \text{ g}} \times 100 = 1.481\% \text{ H}$

$\% P = \dfrac{30.97 \text{ g P}}{136.09 \text{ g}} \times 100 = 22.76\% \text{ P}$

$\% O = \dfrac{64.00 \text{ g O}}{136.09 \text{ g}} \times 100 = 47.03\% \text{ O}$

h. mass of K present = 3 (39.10) g = 117.3 g
mass of P present = 30.97 g = 30.97 g
molar mass of K_3P = 148.27 g = 148.3 g

$\% K = \dfrac{117.3 \text{ g K}}{148.3 \text{ g}} \times 100 = 79.10\% \text{ K}$

$\% P = \dfrac{30.97 \text{ g P}}{148.3 \text{ g}} \times 100 = 20.88\% \text{ P}$

28. a. molar mass of $CuBr_2$ = 223.4 g

$\% Cu = \dfrac{63.55 \text{ g Cu}}{223.4 \text{ g}} \times 100 = 28.45\% \text{ Cu}$

b. molar mass of CuBr = 143.5 g

$\% Cu = \dfrac{63.55 \text{ g Cu}}{143.5 \text{ g}} \times 100 = 44.29\% \text{ Cu}$

c. molar mass of $FeCl_2$ = 126.75 g

$\% Fe = \dfrac{55.85 \text{ g Fe}}{126.75 \text{ g}} \times 100 = 44.06\% \text{ Fe}$

d. molar mass of $FeCl_3$ = 162.2 g

$\% Fe = \dfrac{55.85 \text{ g Fe}}{162.2 \text{ g}} \times 100 = 34.43\% \text{ Fe}$

e. molar mass of CoI_2 = 312.7 g

$\% Co = \dfrac{58.93 \text{ g Co}}{312.7 \text{ g}} \times 100 = 18.85\% \text{ Co}$

f. molar mass of CoI_3 = 439.6 g

$\% Co = \dfrac{58.93 \text{ g Co}}{439.6 \text{ g}} \times 100 = 13.41\% \text{ Co}$

Chemical Composition

g.   molar mass of SnO = 134.7 g

$$\% \text{ Sn} = \frac{118.7 \text{ g Sn}}{134.7 \text{ g}} \times 100 = 88.12\% \text{ Sn}$$

h.   molar mass of SnO_2 = 150.7 g

$$\% \text{ Sn} = \frac{118.7 \text{ g Sn}}{150.7 \text{ g}} \times 100 = 78.77\% \text{ Sn}$$

29. a.   molar mass of $C_6H_{10}O_4$ = 146.1 g

$$\% \text{ C} = \frac{72.06 \text{ g C}}{146.1 \text{ g}} \times 100 = 49.32\% \text{ C}$$

b.   molar mass of NH_4NO_3 = 80.05 g

$$\% \text{ N} = \frac{28.02 \text{ g N}}{80.05 \text{ g}} \times 100 = 35.00\% \text{ N}$$

c.   molar mass of $C_8H_{10}N_4O_2$ = 194.2 g

$$\% \text{ C} = \frac{96.08 \text{ g C}}{194.2 \text{ g}} \times 100 = 49.47\% \text{ C}$$

d.   molar mass of ClO_2 = 67.45 g

$$\% \text{ Cl} = \frac{35.45 \text{ g Cl}}{67.45 \text{ g}} \times 100 = 52.56\% \text{ Cl}$$

e.   molar mass of $C_6H_{11}OH$ = 100.2 g

$$\% \text{ C} = \frac{72.06 \text{ g C}}{100.2 \text{ g}} \times 100 = 71.92\% \text{ C}$$

f.   molar mass of $C_6H_{12}O_6$ = 180.2 g

$$\% \text{ C} = \frac{72.06 \text{ g C}}{180.2 \text{ g}} \times 100 = 39.99\% \text{ C}$$

g.   molar mass of $C_{20}H_{42}$ = 282.5 g

$$\% \text{ C} = \frac{240.2 \text{ g C}}{282.5 \text{ g}} \times 100 = 85.03\% \text{ C}$$

h.   molar mass of C_2H_5OH = 46.07 g

$$\% \text{ C} = \frac{24.02 \text{ g C}}{46.07 \text{ g}} \times 100 = 52.14\% \text{ C}$$

30. a.   molar mass of NH_4Cl = 53.49 g

molar mass of NH_4^+ ion = 18.04 g

Copyright © Houghton Mifflin Company. All rights reserved.

$$\% \; NH_4^+ = \frac{18.04 \; g \; NH_4^+}{53.49 \; g} \times 100 = 33.73\% \; NH_4^+$$

b.   molar mass of $CuSO_4$ = 159.62

molar mass of Cu^{2+} ion = 63.55 g

$$\% \; Cu^{2+} = \frac{63.55 \; g \; Cu^{2+}}{159.62 \; g} \times 100 = 39.81\% \; Cu^{2+}$$

c.   molar mass of $AuCl_3$ = 303.4 g

molar mass of Au^{3+} ion = 197.0 g

$$\% \; Au^{3+} = \frac{197.0 \; g \; Au^{3+}}{303.4 \; g} \times 100 = 64.93\% \; Au^{3+}$$

d.   molar mass of $AgNO_3$ = 169.9 g

molar mass of Ag^+ ion = 107.9 g

$$\% \; Ag^+ = \frac{107.9 \; g \; Ag^+}{169.9 \; g} \times 100 = 63.51\% \; Ag^+$$

31. To determine the *empirical* formula of a new compound, the composition of the compound by mass must be known. To determine the *molecular* formula of the compound, the molar mass of the compound must also be known.

32. The empirical formula represents the smallest whole number ratio of the elements present in a compound. The molecular formula indicates the actual number of atoms of each element found in a molecule of the substance.

33.   a.   NaO

       b. $C_4H_3O_2$

       c. $C_{12}H_{12}N_2O_3$ is already the empirical formula

       d. C_2H_3Cl

34.   a.   yes (each of these has empirical formula CH)

       b.   no (the number of hydrogen atoms is wrong)

       c.   yes (both have empirical formula NO_2)

       d.   no (the number of hydrogen and oxygen atoms is wrong)

35. $0.1929 \; g \; C \times \dfrac{1 \; mol \; C}{12.01 \; g \; C} = 0.01606 \; mol \; C$

$0.01079 \; g \; H \times \dfrac{1 \; mol \; H}{1.008 \; g \; H} = 0.01070 \; mol \; H$

Chemical Composition

$$0.08566 \text{ g O} \times \frac{1 \text{ mol O}}{16.00 \text{ g O}} = 0.005354 \text{ mol O}$$

$$0.1898 \text{ g Cl} \times \frac{1 \text{ mol Cl}}{35.45 \text{ g Cl}} = 0.005354 \text{ mol Cl}$$

Dividing each number of moles by the smallest number of moles gives

$$\frac{0.01606 \text{ mol C}}{0.005354} = 3.000 \text{ mol C}$$

$$\frac{0.01070 \text{ mol H}}{0.005354} = 1.999 \text{ mol H}$$

$$\frac{0.005354 \text{ mol O}}{0.005354} = 1.000 \text{ mol O}$$

$$\frac{0.005354 \text{ Cl}}{0.005354} = 1.000 \text{ mol Cl}$$

The empirical formula is C_3H_2OCl

36. $$2.514 \text{ g Ca} \times \frac{1 \text{ mol}}{40.08 \text{ g Ca}} = 0.06272 \text{ mol Ca}$$

The increase in mass represents the oxygen with which the calcium reacted:

$$1.004 \text{ g O} \times \frac{1 \text{ mol O}}{16.00 \text{ g O}} = 0.06275 \text{ mol O}$$

Since we have effectively the same number of moles of Ca and O, the empirical formula must be CaO.

37. Consider having 100.0 g of the compound. Then the percentages of the elements present are numerically equal to their masses in grams.

$$58.84 \text{ g Ba} \times \frac{1 \text{ mol}}{137.3 \text{ g Ba}} = 0.4286 \text{ mol Ba}$$

$$13.74 \text{ g S} \times \frac{1 \text{ mol}}{32.07 \text{ g S}} = 0.4284 \text{ mol S}$$

$$27.43 \text{ g O} \times \frac{1 \text{ mol O}}{16.00 \text{ g O}} = 1.714 \text{ mol O}$$

Dividing each number of moles by the smallest number of moles (0.4284 mol S) gives

$$\frac{0.4286 \text{ mol Ba}}{0.4284} = 1.000 \text{ mol Ba}$$

$$\frac{0.4284 \text{ mol S}}{0.4284} = 1.000 \text{ mol S}$$

$$\frac{1.714 \text{ mol O}}{0.4284} = 4.001 \text{ mol O}$$

The empirical formula is BaSO₄.

38. The mass of chlorine involved in the reaction is 6.280 – 1.271 = 5.009 g Cl

$$1.271 \text{ g Al} \times \frac{1 \text{ mol Al}}{26.98 \text{ g Al}} = 0.04711 \text{ mol Al}$$

$$5.009 \text{ g Cl} \times \frac{1 \text{ mol Cl}}{34.45 \text{ g Cl}} = 0.1413 \text{ mol Cl}$$

Dividing each of these numbers of moles by the smaller (0.04711 mol Al) shows that the empirical formula is AlCl₃.

39. Consider 100.0 g of the compound.

$$55.06 \text{ g Co} \times \frac{1 \text{ mol}}{58.93 \text{ g Co}} = 0.9343 \text{ mol Co}$$

If the sulfide of cobalt is 55.06% Co, then it is 44.94% S by mass.

$$44.94 \text{ g S} \times \frac{1 \text{ mol}}{32.07 \text{ g S}} = 1.401 \text{ mol S}$$

Dividing each number of moles by the smaller (0.9343 mol Co) gives

$$\frac{0.9343 \text{ mol Co}}{0.9343} = 1.000 \text{ mol Co}$$

$$\frac{1.401 \text{ mol S}}{0.9343} = 1.500 \text{ mol S}$$

Multiplying by two, to convert to whole numbers of moles, gives the empirical formula for the compound as Co₂S₃.

40. $$2.461 \text{ g Ca} \times \frac{1 \text{ mol Ca}}{40.08 \text{ g Ca}} = 0.06140 \text{ mol Ca}$$

$$4.353 \text{ g Cl} \times \frac{1 \text{ mol Cl}}{35.45 \text{ g Cl}} = 0.1228 \text{ mol Cl}$$

Dividing each of the number of moles by the smaller (0.06140 mol Ca) shows that the empirical formula is CaCl₂.

41. $$10.00 \text{ g Cu} \times \frac{1 \text{ mol}}{63.55 \text{ g Cu}} = 0.1574 \text{ mol Cu}$$

$$2.52 \text{ g O} \times \frac{1 \text{ mol O}}{16.00 \text{ g O}} = 0.158 \text{ mol O}$$

The numbers are almost equal: the empirical formula is CuO.

Chemical Composition

42. Consider 100.0 g of the compound.

    $$33.88 \text{ g Cu} \times \frac{1 \text{ mol Cu}}{63.55 \text{ g Cu}} = 0.5331 \text{ mol Cu}$$

    $$14.94 \text{ g N} \times \frac{1 \text{ mol N}}{14.01 \text{ g N}} = 1.066 \text{ mol N}$$

    $$51.18 \text{ g} \times \frac{1 \text{ mol O}}{16.00 \text{ g O}} = 3.199 \text{ mol O}$$

    Dividing each number of moles by the smaller number of moles (0.5331 mol Cu) gives

    $$\frac{0.5331 \text{ mol Cu}}{0.5331} = 1.000 \text{ mol Cu}$$

    $$\frac{1.066 \text{ mol N}}{0.5331} = 2.000 \text{ mol N}$$

    $$\frac{3.199 \text{ mol O}}{0.5331} = 6.001 \text{ mol O}$$

    The empirical formula is CuN_2O_6 [i.e., $Cu(NO_3)_2$]

43. Compound 1: Assume 100.0 g of the compound.

    $$83.12 \text{ g Na} \times \frac{1 \text{ mol Na}}{22.99 \text{ g Na}} = 3.615 \text{ mol Na}$$

    $$16.88 \text{ g N} \times \frac{1 \text{ mol N}}{14.01 \text{ g N}} = 1.205 \text{ mol Na}$$

    Dividing each number of moles by the smaller (1.205 mol Na) indicates that the formula of Compound 1 is Na_3N.

    Compound 2: Assume 100.0 g of the compound.

    $$35.36 \text{ g Na} \times \frac{1 \text{ mol Na}}{22.99 \text{ g Na}} = 1.538 \text{ mol Na}$$

    $$64.64 \text{ g N} \times \frac{1 \text{ mol N}}{14.01 \text{ g N}} = 4.614 \text{ mol N}$$

    Dividing each number of moles by the smaller (1.538 mol Na) indicates that the formula of Compound 2 is NaN_3.

44. The *empirical formula* of a compound represents only the smallest whole number relationship between the number and type of atoms in a compound, whereas the *molecular formula* represents the actual number of atoms of each type in a true molecule of the substance. Many compounds (for example, H_2O) may have the same empirical and molecular formulas.

45. If only the empirical formula is known, the molar mass of the substance must be determined before the molecular formula can be calculated.

46. empirical formula mass of CH_2O = 30 g

$$n = \frac{\text{molar mass}}{\text{empirical formula mass}} = \frac{90 \text{ g}}{30 \text{ g}} = 3$$

molecular formula is $(CH_2O)_3 = C_3H_6O_3$.

47. empirical formula mass of CH_2 = 14

$$n = \frac{\text{molar mass}}{\text{empirical formula mass}} = \frac{84 \text{ g}}{14 \text{ g}} = 6$$

molecular formula is $(CH_2)_6 = C_6H_{12}$.

48. empirical formula mass of CH_4O = 32.04 g

$$n = \frac{\text{molar mass}}{\text{empirical formula mass}} = \frac{192 \text{ g}}{32.04 \text{ g}} = 6$$

molecular formula is $(CH_4O)_6 = C_6H_{24}O_6$.

49. Consider 100.0 g of the compound.

$$42.87 \text{ g C} \times \frac{1 \text{ mol C}}{12.01 \text{ g C}} = 3.570 \text{ mol C}$$

$$3.598 \text{ g H} \times \frac{1 \text{ mol H}}{1.008 \text{ g H}} = 3.569 \text{ mol H}$$

$$28.55 \text{ g O} \times \frac{1 \text{ mol O}}{16.00 \text{ g O}} = 1.784 \text{ mol O}$$

$$25.00 \text{ g N} \times \frac{1 \text{ mol N}}{14.01 \text{ g N}} = 1.784 \text{ mol N}$$

Dividing each number of moles by the smallest number of moles (1.784 mol O) gives

$$\frac{3.570 \text{ mol C}}{1.784} = 2.001 \text{ mol C}$$

$$\frac{3.569 \text{ mol H}}{1.784} = 2.001 \text{ mol H}$$

$$\frac{1.784 \text{ mol O}}{1.784} = 1.000 \text{ mol O}$$

$$\frac{1.784 \text{ mol N}}{1.784} = 1.000 \text{ mol N}$$

The empirical formula of the compound is C_2H_2ON, empirical formula mass of C_2H_2ON = 56

$$n = \frac{\text{molar mass}}{\text{empirical formula mass}} = \frac{168 \text{ g}}{56 \text{ g}} = 3$$

The molecular formula is $(C_2H_2ON)_3 = C_6H_6O_3N_3$.

Chemical Composition

50. Consider 100.0 g of the compound.

$$65.45 \text{ g C} \times \frac{1 \text{ mol C}}{12.01 \text{ g C}} = 5.450 \text{ mol C}$$

$$5.492 \text{ g H} \times \frac{1 \text{ mol H}}{1.008 \text{ g H}} = 5.448 \text{ mol H}$$

$$29.06 \text{ g O} \times \frac{1 \text{ mol O}}{16.00 \text{ g O}} = 1.816 \text{ mol O}$$

Dividing each number of moles by the smallest number of moles (1.816 mol O) gives

$$\frac{5.450 \text{ mol C}}{1.816} = 3.001 \text{ mol C}$$

$$\frac{5.448 \text{ mol H}}{1.816} = 3.000 \text{ mol H}$$

$$\frac{1.816 \text{ mol O}}{1.816} = 1.000 \text{ mol O}$$

The empirical formula is C_3H_3O, and the empirical formula mass is approximately 55 g.

$$n = \frac{\text{molar mass}}{\text{empirical formula mass}} = \frac{110 \text{ g}}{55 \text{ g}} = 2$$

The molecular formula is $(C_3H_3O)_2 = C_6H_6O_2$.

51. [1] c   [6] d
     [2] e   [7] a
     [3] j   [8] g
     [4] h   [9] i
     [5] b   [10] f

52.

5.00 g Al	0.185 mol	1.12×10^{23} atoms
0.140 g Fe	0.00250 mol	1.51×10^{21} atoms
2.7×10^2 g Cu	4.3 mol	2.6×10^{24} atoms
0.00250 g Mg	1.03×10^{-4} mol	6.19×10^{19} atoms
0.062 g Na	2.7×10^{-3} mol	1.6×10^{21} atoms
3.95×10^{-18} g U	1.66×10^{-20} mol	1.00×10^4 atoms

53.

4.24 g	0.0543 mol	3.27×10^{22} molec.	3.92×10^{23} atoms
4.04 g	0.224 mol	1.35×10^{23} molec.	4.05×10^{23} atoms
1.98 g	0.0450 mol	2.71×10^{22} molec.	8.13×10^{22} atoms
45.9 g	1.26 mol	7.59×10^{23} molec.	1.52×10^{24} atoms

| 126 g | 6.99 mol | 4.21 x 10^{24} molec. | 1.26 x 10^{25} atoms |
| 0.297 g | 0.00927 mol | 5.58 x 10^{21} molec. | 3.35 x 10^{22} atoms |

54. magnesium/nitrogen compound:

mass of nitrogen contained = 1.2791 g – 0.9240 g = 0.3551 g N

$$0.9240 \text{ g Mg} \times \frac{1 \text{ mol Mg}}{24.31 \text{ g Mg}} = 0.03801 \text{ mol Mg}$$

$$0.3551 \text{ g N} \times \frac{1 \text{ mol N}}{14.01 \text{ g N}} = 0.02535 \text{ mol N}$$

Dividing each number of moles by the smaller number of moles gives

$$\frac{0.03801 \text{ mol Mg}}{0.02535} = 1.499 \text{ mol Mg}$$

$$\frac{0.02535 \text{ mol N}}{0.02535} = 1.000 \text{ mol N}$$

Multiplying by two, to convert to whole numbers, gives the empirical formula as Mg_3N_2.

magnesium/oxygen compound:

Consider 100.0 g of this compound.

$$60.31 \text{ g Mg} \times \frac{1 \text{ mol Mg}}{24.31 \text{ g Mg}} = 2.481 \text{ mol Mg}$$

$$39.69 \text{ g O} \times \frac{1 \text{ mol O}}{16.00 \text{ g O}} = 2.481 \text{ mol O}$$

Since the numbers of moles are the same, the compound contains the same relative number of Mg and O atoms: the empirical formula is MgO.

55. For the first compound (*restricted* amount of oxygen)

$$2.118 \text{ g Cu} \times \frac{1 \text{ mol Cu}}{63.54 \text{ g Cu}} = 0.03333 \text{ mol Cu}$$

$$0.2666 \text{ g O} \times \frac{1 \text{ mol O}}{16.00 \text{ g O}} = 0.01666 \text{ mol O}$$

Since the number of moles of Cu (0.03333 mol) is twice the number of moles of O (0.01666 mol), the empirical formula is Cu_2O.

For the second compound (stream of pure oxygen)

$$2.118 \text{ g Cu} \times \frac{1 \text{ mol Cu}}{63.54 \text{ g Cu}} = 0.03333 \text{ mol Cu}$$

$$0.5332 \text{ g O} \times \frac{1 \text{ mol O}}{16.00 \text{ g O}} = 0.03333 \text{ mol O}$$

Since the numbers of moles are the same, the empirical formula is CuO.

Chemical Composition

56.
[1]	g	[6]	I	
[2]	c	[7]	f	
[3]	b	[8]	h	
[4]	a	[9]	e	
[5]	j	[10]	d	

57. $2.24 \text{ g Co} \times \dfrac{55.85 \text{ g Fe}}{58.93 \text{ g Co}} = 2.12 \text{ g Fe}$

58. $2.24 \text{ g Fe} \times \dfrac{58.93 \text{ g Co}}{55.85 \text{ g Fe}} = 2.36 \text{ g Co}$

59. Consider 100.0 g of the compound.

$25.45 \text{ g Cu} \times \dfrac{1 \text{ mol Cu}}{63.55 \text{ g Cu}} = 0.4005 \text{ mol Cu}$

$12.84 \text{ g S} \times \dfrac{1 \text{ mol S}}{32.07 \text{ g S}} = 0.4004 \text{ mol S}$

$4.036 \text{ g H} \times \dfrac{1 \text{ mol H}}{1.008 \text{ g H}} = 4.004 \text{ mol H}$

$57.67 \text{ g O} \times \dfrac{1 \text{ mol O}}{16.00 \text{ g O}} = 3.604 \text{ mol O}$

Dividing each number of moles by the smallest number of moles gives

$\dfrac{0.4005 \text{ mol Cu}}{0.4004} = 1.000 \text{ mol Cu}$

$\dfrac{0.4004 \text{ mol S}}{0.4004} = 1.000 \text{ mol S}$

$\dfrac{4.004 \text{ mol H}}{0.4004} = 10.00 \text{ mol H}$

$\dfrac{3.604 \text{ mol O}}{0.4004} = 9.001 \text{ mol O}$

The empirical formula is $CuSH_{10}O_9$ (which is usually written as $CuSO_4 \cdot 5H_2O$).

60. $0.2990 \text{ g C} \times \dfrac{1 \text{ mol C}}{12.01 \text{ g C}} = 0.02490 \text{ mol C}$

$0.05849 \text{ g H} \times \dfrac{1 \text{ mol H}}{1.008 \text{ g H}} = 0.05803 \text{ mol H}$

$0.2318 \text{ g N} \times \dfrac{1 \text{ mol N}}{14.01 \text{ g N}} = 0.01655 \text{ mol N}$

$0.1328 \text{ g O} \times \dfrac{1 \text{ mol O}}{16.00 \text{ g O}} = 0.008300 \text{ mol O}$

Dividing each number of moles by the smallest number of moles (0.008300 mol O) gives

$$\frac{0.02490 \text{ mol C}}{0.008300} = 3.000 \text{ mol C}$$

$$\frac{0.05803 \text{ mol H}}{0.008300} = 6.992 \text{ mol H}$$

$$\frac{0.01655 \text{ mol N}}{0.008300} = 1.994 \text{ mol N}$$

$$\frac{0.008300 \text{ mol O}}{0.008300} = 1.000 \text{ mol O}$$

The empirical formula is $C_3H_7N_2O$.

61. Mass of oxygen in compound = 4.33 g – 4.01 g = 0.32 g O

$$4.01 \text{ g Hg} \times \frac{1 \text{ mol Hg}}{200.6 \text{ g Hg}} = 0.0200 \text{ mol Hg}$$

$$0.32 \text{ g O} \times \frac{1 \text{ mol O}}{16.00 \text{ g O}} = 0.020 \text{ mol O}$$

Since the numbers of moles are equal, the empirical formula is HgO.

62. Assume we have 100.0 g of the compound.

$$65.95 \text{ g Ba} \times \frac{1 \text{ mol Ba}}{137.3 \text{ g Ba}} = 0.4803 \text{ mol Ba}$$

$$34.05 \text{ g Cl} \times \frac{1 \text{ mol Cl}}{35.45 \text{ g Cl}} = 0.9605 \text{ mol Cl}$$

Dividing each of these number of moles by the smaller number gives

$$\frac{0.4803 \text{ mol Ba}}{0.4803} = 1.000 \text{ mol Ba}$$

$$\frac{0.9605 \text{ mol Cl}}{0.4803} = 2.000 \text{ mol Cl}$$

The empirical formula is then $BaCl_2$.

63. We need to find the subscripts for $H_xN_yO_z$, where $x:y:z$ is the mole ratio of H:N:O atoms. We are given that $x = 4.0$ moles H.
To solve for y, we use

$$56.0 \text{ g N} \times \frac{1 \text{ mol N}}{14.01 \text{ g N}} = 4.00 \text{ moles N}$$

Chemical Composition

To solve for z, we use

$$7.2 \times 10^{24} \text{ atoms O} \times \frac{1 \text{ mol O}}{6.022 \times 10^{23} \text{ atoms}} = 12 \text{ moles O}$$

The mole ratio is 4:4:12 or 1:1:3. The empirical formula is HNO_3.

64. For every 100.0 g of A_2O, we have 63.7 g A and 36.3 g O.

$$36.3 \text{ g O} \times \frac{1 \text{ mol O}}{16.00 \text{ g O}} = 2.27 \text{ mol O}$$

The ratio between A and O is 2:1; with 2.27 mol O we must have 4.54 mol A.

$$\frac{63.7 \text{ g A}}{4.54 \text{ mol A}} = 14.0 \text{ g/mol}$$

Thus, A must be nitrogen. The compound is N_2O.

65. We need to find the subscripts for $C_aH_bO_cS_d$, where $a:b:c:d$ is the mole ratio of C:H:O:S atoms. We are given the following relationships:

$b = 2a$

$a = c$

$b = 8d$

We can see that d will be the smallest subscript. Thus, let $d = x$ and we get

$b = 8x$

$2a = 8x$ or $\quad a = 4x$

since $a = c$, $\quad c = 4x$

Thus, we have $C_{4x}H_{8x}O_{4x}S_x$. The empirical formula is $C_4H_8O_4S$, which has a molar mass of about 152 g/mol [4(12.01) + 8(1.008) + 4(16.00) + 1(32.07)]. We are given that the molar mass of the compound is 152 g/mol. Thus, the molecular formula is $C_4H_8O_4S$.

Chapter 7 Chemical Reactions: An Introduction

1. The types of evidence for a chemical reaction mentioned in the text are: a change in color, formation of a solid, evolution of a gas, and absorption or evolution of heat. Other bits of evidence that might also be observed include appearance or disappearance of a characteristic odor, or separation of the reaction mixture into layers of visibly different composition.

2. The fact that there is a decrease in mass is the best evidence for reaction. If mass has been lost, then it is likely that a gaseous substance, which has escaped into the environment has been produced by the heating. The fact that the chalk crumbles into a powder may be taken as secondary evidence that the chalk has been converted into something which does not stick together well.

3. The fact that the material in the drain, which did not dissolve in water, dissolves when the drain cleaner is added, suggests that rather than simple dissolving, the material in the drain has undergone a chemical change which makes it soluble. You may also have noticed that the drain cleaner evolved *heat* when added to the drain: evolution or absorption of heat is also often a sign of a chemical reaction.

4. The observation that the odor of the aspirin has changed to the odor of vinegar indicates that a new substance has been produced. This is a chemical reaction.

5. reactants, products

6. atoms

7. gaseous

8. water

9. $CaCO_3(s) \rightarrow CaO(s) + CO_2(g)$

10. $C_3H_8(g) + O_2(g) \rightarrow CO_2(g) + H_2O(g)$

11. $H_2(g) + O_2(g) \rightarrow H_2O(g)$

12. $(NH_4)_2CO_3(s) \rightarrow NH_3(g) + CO_2(g) + H_2O(g)$

13. $Ag_2O(s) \rightarrow Ag(s) + O_2(g)$

14. $CO(g) + H_2(g) \rightarrow CH_3OH(l)$

15. $B_2O_3(s) + Mg(s) \rightarrow B(s) + MgO(s)$

16. $Ca(s) + H_2O(l) \rightarrow Ca(OH)_2(s) + H_2(g)$

17. $P_4(s) + Cl_2(g) \rightarrow PCl_3(s)$

18. $Mg(OH)_2(s) + HCl(aq) \rightarrow MgCl_2(aq) + H_2O(l)$

Chemical Reactions: An Introduction

19. $NH_4NO_3(s) \rightarrow N_2O(g) + H_2O(g)$

20. $H_2S(g) + O_2(g) \rightarrow SO_2(g) + H_2O(g)$

21. $C_2H_2(g) + O_2(g) \rightarrow CO_2(g) + H_2O(g)$

22. $Fe_2O_3(s) + CO(g) \rightarrow Fe(s) + CO_2(g)$

23. $BaO(s) + Al(s) \rightarrow Ba(s) + Al_2O_3(s)$

    $CaO(s) + Al(s) \rightarrow Ca(s) + Al_2O_3(s)$

    $SrO(s) + Al(s) \rightarrow Sr(s) + Al_2O_3(s)$

24. $O_2(g) \rightarrow O_3(g)$

25. $CH_4(g) + Cl_2(g) \rightarrow CCl_4(l) + HCl(g)$

26. $NH_3(g) + HNO_3(aq) \rightarrow NH_4NO_3(s)$

27. $PbS(s) + O_2(g) \rightarrow PbO(s) + SO_2(g)$

    $PbO(s) + C(s) \rightarrow Pb(l) + CO_2(g)$

28. $Xe(g) + F_2(g) \rightarrow XeF_4(s)$

29. $NH_4NO_3(s) \rightarrow N_2(g) + O_2(g) + H_2O(g)$

30. $Ag(s) + HNO_3(aq) \rightarrow AgNO_3(aq) + H_2(g)$

31. formula

32. whole numbers

33. a. $H_2O_2 \rightarrow H_2O + O_2$

    Balance oxygen: $2H_2O_2 \rightarrow 2H_2O + O_2$

    Balanced equation: $2H_2O_2(aq) \rightarrow 2H_2O(l) + O_2(g)$

    b. $Ag + H_2S \rightarrow Ag_2S + H_2$

    Balance silver: $2Ag + H_2S \rightarrow Ag_2S + H_2$

    Balanced equation: $2Ag(s) + H_2S(g) \rightarrow Ag_2S(s) + H_2(g)$

    c. $FeO + C \rightarrow Fe + CO_2$

    Balance oxygen: $2FeO + C \rightarrow Fe + CO_2$

    Balance iron: $2FeO + C \rightarrow 2Fe + CO_2$

    Balanced equation: $2FeO(s) + C(s) \rightarrow 2Fe(l) + CO_2(g)$

Copyright © Houghton Mifflin Company. All rights reserved.

d. $Cl_2 + KI \rightarrow KCl + I_2$

    Balance chlorine: $Cl_2 + KI \rightarrow \mathbf{2}KCl + I_2$

    Balance iodine: $Cl_2 + \mathbf{2}KI \rightarrow 2KCl + I_2$

    Balanced equation: $Cl_2(g) + 2KI(aq) \rightarrow 2KCl(aq) + I_2(s)$

34. a. $CaF_2 + H_2SO_4 \rightarrow CaSO_4 + HF$

    Balance fluorine: $CaF_2 + H_2SO_4 \rightarrow CaSO_4 + \mathbf{2}HF$

    Balanced equation: $CaF_2(s) + H_2SO_4(l) \rightarrow CaSO_4(s) + 2HF(g)$

b. $KBr + H_3PO_4 \rightarrow K_3PO_4 + HBr$

    Balance potassium: $\mathbf{3}KBr + H_3PO_4 \rightarrow K_3PO_4 + HBr$

    Balance bromine: $3KBr + H_3PO_4 \rightarrow K_3PO_4 + \mathbf{3}HBr$

    Balanced equation: $3KBr(s) + H_3PO_4(aq) \rightarrow K_3PO_4(aq) + 3HBr(g)$

c. $TiCl_4 + Na \rightarrow NaCl + Ti$

    Balance chlorine: $TiCl_4 + Na \rightarrow \mathbf{4}NaCl + Ti$

    Balance sodium: $TiCl_4 + \mathbf{4}Na \rightarrow 4NaCl + Ti$

    Balanced equation: $TiCl_4(l) + 4Na(s) \rightarrow 4NaCl(s) + Ti(s)$

d. $K_2CO_3 \rightarrow K_2O + CO_2$ This equation is already balanced!

35. a. $SiI_4(s) + 2Mg(s) \rightarrow Si(s) + 2MgI_2(s)$

b. $MnO_2(s) + 2Mg(s) \rightarrow Mn(s) + 2MgO(s)$

c. $8Ba(s) + S_8(s) \rightarrow 8BaS(s)$

d. $4NH_3(g) + 3Cl_2(g) \rightarrow 3NH_4Cl(s) + NCl_3(g)$

36. a. $Ba(NO_3)_2(aq) + Na_2CrO_4(aq) \rightarrow BaCrO_4(s) + 2NaNO_3(aq)$

b. $PbCl_2(aq) + K_2SO_4(aq) \rightarrow PbSO_4(s) + 2KCl(aq)$

c. $C_2H_5OH(l) + 3O_2(g) \rightarrow 2CO_2(g) + 3H_2O(l)$

d. $CaC_2(s) + 2H_2O(l) \rightarrow Ca(OH)_2(s) + C_2H_2(g)$

37. $Al(s) + O_2(g) \rightarrow Al_2O_3(s)$

38. $KNO_3(s) + C(s) \rightarrow K_2CO_3(s) + CO(g) + N_2(g)$

39. $C_{12}H_{22}O_{11}(aq) + H_2O(l) \rightarrow 4C_2H_5OH(aq) + 4CO_2(g)$

40. $2H_2(g) + CO(g) \rightarrow CH_3OH(l)$

41. $2Al_2O_3(s) + 3C(s) \rightarrow 4Al(s) + 3CO_2(g)$

Chemical Reactions: An Introduction

42. $Fe_3O_4(s) + 4H_2(g) \rightarrow 3Fe(s) + 4H_2O(g)$

    $Fe_3O_4(s) + 4CO(g) \rightarrow 3Fe(s) + 4CO_2(g)$

43. $2Li(s) + S(s) \rightarrow Li_2S(s)$

    $2Na(s) + S(s) \rightarrow Na_2S(s)$

    $2K(s) + S(s) \rightarrow K_2S(s)$

    $2Rb(s) + S(s) \rightarrow Rb_2S(s)$

    $2Cs(s) + S(s) \rightarrow Cs_2S(s)$

    $2Fr(s) + S(s) \rightarrow Fr_2S(s)$

44. $2KClO_3(s) \rightarrow 2KCl(s) + 3O_2(g)$

45. $2H_2O_2(aq) \rightarrow 2H_2O(g) + O_2(g)$

46. $CaSiO_3(s) + 6HF(g) \rightarrow CaF_2(aq) + SiF_4(g) + 3H_2O(l)$

47. Many over-the-counter antacids contain either carbonate ion (CO_3^{2-}) or hydrogen carbonate ion (HCO_3^-). When either of these encounter stomach acid (primarily HCl), carbon dioxide gas is released.

48. $K_2CrO_4(aq) + BaCl_2(aq) \rightarrow BaCrO_4(s) + 2KCl(aq)$

49. $Mg(s) + O_2(g) \rightarrow MgO(s)$

50. $CuO(s) + H_2SO_4(aq) \rightarrow CuSO_4(aq) + H_2O(l)$

51. $PbS(s) + O_2(g) \rightarrow PbO(s) + SO_2(g)$

52. We are considering the equation

    $C_6H_6 + HNO_3 \rightarrow C_aH_bN_cO_d + H_2O$

    For every 100.0 g of the compound we have

    $33.8 \text{ g C} \times \dfrac{1 \text{ mol C}}{12.01 \text{ g C}} = 2.81 \text{ mol C}$

    $1.42 \text{ g H} \times \dfrac{1 \text{ mol H}}{1.008 \text{ g H}} = 1.41 \text{ mol H}$

    $19.7 \text{ g N} \times \dfrac{1 \text{ mol N}}{14.01 \text{ g N}} = 1.41 \text{ mol N}$

    $45.1 \text{ g O} \times \dfrac{1 \text{ mol O}}{16.00 \text{ g O}} = 2.82 \text{ mol O}$

Reducing this mole ratio to smallest whole numbers, we get an empirical formula of C_2HNO_2, which has a molar mass of about 71 g/mol

$$[2(12.01) + 1(1.008) + 1(14.01) + 2(16.00)]$$

We are told that the molar mass of the compound is 213 g/mol, which is 3 times as great as the molar mass of the empirical formula. This makes the molecular formula $C_6H_3N_3O_6$. The balanced equation is therefore

$$C_6H_6(l) + 3HNO_3(aq) \rightarrow C_6H_3N_3O_6(l) + 3H_2O(l)$$

Chapter 8 Reactions in Aqueous Solutions

1. Water is the most universal of all liquids. Water has a relatively large heat capacity, and a relatively large liquid range, which means it can absorb the heat liberated by many reactions while still remaining in the liquid state. Water is very polar and dissolves well both ionic solutes and solutes with which it can hydrogen bond (this is especially important to the biochemical reactions of the living cell).

2. Driving forces are types of *changes* in a system which pull a reaction in the *direction of product formation*; driving forces include: formation of a *solid*, formation of *water*, formation of a *gas*, transfer of electrons.

3. The net charge of a precipitate must be *zero*. The total number of positive charges equals the total number of negative charges.

4. When an electrolyte such as NaCl (sodium chloride) is dissolved in water, the resulting solution consists of separate, individual, discrete sodium ions (Na^+) and separate, individual, discrete chloride ions (Cl^-). There are no identifiable NaCl units in such a solution.

5. A substance is said to be a strong electrolyte if *each* unit of the substance produces separated, distinct ions when the substance is dissolved in water. NaCl and KNO_3 are both strong electrolytes.

6. $NaNO_3$ must be soluble in water.

7. For most practical purposes, "insoluble" and "slightly" soluble mean the same thing. The difference between "insoluble" and "slightly soluble" could be crucial if, for example, a substance were highly toxic and were found in a water supply.

8.
   a. soluble (Rule 1: most nitrate salts are soluble)
   b. soluble (Rule 2: most potassium salts are soluble)
   c. soluble (Rule 2: most sodium salts are soluble)
   d. insoluble (Rule 5: most hydroxide compounds are insoluble)
   e. insoluble (Rule 3: exception for chloride salts)
   f. soluble (Rule 2: most ammonium salts are soluble)
   g. insoluble (Rule 6: most sulfide salts are insoluble)
   h. insoluble (Rule 4: exception for sulfate salts)

9.
   a. Rule 5: most hydroxides are only slightly soluble
   b. Rule 6: most carbonates are only slightly soluble
   c. Rule 6: most phosphates are only slightly soluble
   d. Rule 3: exception to the rule for chlorides

10.
   a. CaSO₄. Rule 4: exception to the rule for sulfates

   b. AgI. Rule 3: although the text does not mention it explicitly, as you might expect from your knowledge of the periodic table, bromide and iodide compounds of Ag^+, Pb^{2+}, and Hg_2^{2+} are insoluble

   c. $Pb_3(PO_4)_2$. Rule 6: most phosphate salts are only slightly soluble

   d. $Fe(OH)_3$. Rule 5: most hydroxides are only slightly soluble

   e. no precipitate is likely: rules 1, 2, and 4

   f. $BaCO_3$. Rule 6: most carbonate salts are only slightly soluble.

11. The precipitates are marked in boldface type.

   a. No precipitate: $Ba(NO_3)_2$ and HCl are each soluble.

   b. Rule 6: most sulfide salts are insoluble.

   $(NH_4)_2S(aq) + CoCl_2(aq) \rightarrow$ **CoS**$(s) + 2NH_4Cl(aq)$

   c. Rule 4: lead sulfate is a listed exception.

   $H_2SO_4(aq) + Pb(NO_3)_2(aq) \rightarrow$ **PbSO₄**$(s) + 2HNO_3(aq)$

   d. Rule 6: most carbonate salts are insoluble.

   $CaCl_2(aq) + K_2CO_3(aq) \rightarrow$ **CaCO₃**$(s) + 2KCl(aq)$

   e. No precipitate: $NaNO_3$ and $NH_4C_2H_3O_2$ are each soluble.

   f. Rule 6: most phosphate salts are insoluble

   $Na_3PO_4(aq) + CrCl_3(aq) \rightarrow 3NaCl(aq) +$ **CrPO₄**(s)

12. Hint: when balancing equations involving polyatomic ions, especially in precipitation reactions, balance the polyatomic ions as a *unit*, not in terms of the atoms the polyatomic ions contain (e.g., treat nitrate ion, NO_3^-, as a single entity, not as one nitrogen and three oxygen atoms). When finished balancing, however, do be sure to count the individual number of atoms of each type on each side of the equation.

   a. $AgNO_3(aq) + H_2SO_4(aq) \rightarrow Ag_2SO_4(s) + HNO_3(aq)$

   Balance silver: $\mathbf{2}AgNO_3(aq) + H_2SO_4(aq) \rightarrow Ag_2SO_4(s) + HNO_3(aq)$

   Balance nitrate: $2AgNO_3(aq) + H_2SO_4(aq) \rightarrow Ag_2SO_4(s) + \mathbf{2}HNO_3(aq)$

   Balanced equation: $2AgNO_3(aq) + H_2SO_4(aq) \rightarrow Ag_2SO_4(s) + 2HNO_3(aq)$

   b. $Ca(NO_3)_2(aq) + H_2SO_4(aq) \rightarrow CaSO_4(s) + HNO_3(aq)$

   Balance nitrate: $Ca(NO_3)_2(aq) + H_2SO_4(aq) \rightarrow CaSO_4(s) + \mathbf{2}HNO_3(aq)$

   Balanced equation: $Ca(NO_3)_2(aq) + H_2SO_4(aq) \rightarrow CaSO_4(s) + 2HNO_3(aq)$

   c. $Pb(NO_3)_2(aq) + H_2SO_4(aq) \rightarrow PbSO_4(s) + HNO_3(aq)$

   Balance nitrate: $Pb(NO_3)_2(aq) + H_2SO_4(aq) \rightarrow PbSO_4(s) + \mathbf{2}HNO_3(aq)$

   Balanced equation: $Pb(NO_3)_2(aq) + H_2SO_4(aq) \rightarrow PbSO_4(s) + 2HNO_3(aq)$

Reactions in Aqueous Solutions

13. The products are determined by having the ions "switch partners." For example, for a general reaction AB + CD →, the possible products are AD and CB if the ions switch partners. If either AD or CB is insoluble, then a precipitation reaction has occurred. In the following reaction, the formula of the precipitate is given in boldface type.

    a. $(NH_4)_2S(aq) + CoCl_2(aq) \rightarrow$ **$CoS(s)$** $+ 2NH_4Cl(aq)$

    Rule 6: most sulfide salts are only slightly soluble

    b. $FeCl_3(aq) + 3NaOH(aq) \rightarrow$ **$Fe(OH)_3(s)$** $+ 3NaCl(aq)$

    Rule 5: Most hydroxide compounds are only slightly soluble

    c. $CuSO_4(aq) + Na_2CO_3(aq) \rightarrow$ **$CuCO_3(s)$** $+ Na_2SO_4(aq)$

    Rule 6: most carbonate salts are only slightly soluble.

14. The net ionic equation for a reaction indicates *only those ions that go to form the precipitate*, and does not show the spectator ions present in the solutes mixed. The identity of the precipitate is determined from the Solubility Rules.

    a. $Ca^{2+}(aq) + SO_4^{2-}(aq) \rightarrow CaSO_4(s)$

    Rule 4: exception to rule about sulfate salts.

    b. $2Fe^{3+}(aq) + 3CO_3^{2-}(aq) \rightarrow Fe_2(CO_3)_3(s)$

    Rule 6: most carbonate salts are only slightly soluble.

    c. $Ag^+(aq) + I^-(aq) \rightarrow AgI(s)$

    Rule 3: AgI, like AgCl, is insoluble.

    d. $3Co^{2+}(aq) + 3PO_4^{3-}(aq) \rightarrow Co_3(PO_4)_2(s)$

    Rule 6: most phosphate salts are only slightly soluble.

    e. $Hg_2^{2+}(aq) + 2Cl^-(aq) \rightarrow Hg_2Cl_2(s)$

    Rule 3: listed exception to the general rule about chlorides.

    f. $Pb^{2+}(aq) + 2Br^-(aq) \rightarrow PbBr_2(s)$

    Rule 3: like $PbCl_2$, $PbBr_2$ and PbI_2 are also insoluble.

15. $Ag^+(aq) + Cl^-(aq) \rightarrow AgCl(s)$

    $Pb^{2+}(aq) + 2Cl^-(aq) \rightarrow PbCl_2(s)$

    $Hg_2^{2+}(aq) + 2Cl^-(aq) \rightarrow Hg_2Cl_2(s)$

16. $Ca^{2+}(aq) + C_2O_4^{2-}(aq) \rightarrow CaC_2O_4(s)$

17. Strong acids are acids that ionize completely in water. The strong acids are also strong electrolytes.

18. Strong bases are bases that fully produce hydroxide ions when dissolved in water. The strong bases are also strong electrolytes.

19. 1000; 1000

20. $HBr(aq) \rightarrow H^+(aq) + Br^-(aq)$ $HClO_4(aq) \rightarrow H^+(aq) + ClO_4^-(aq)$

21. $RbOH(s) \rightarrow Rb^+(aq) + OH^-(aq)$

    $CsOH(s) \rightarrow Cs^+(aq) + OH^-(aq)$

22. The formulas of the salts are marked in boldface type. Remember that in an acid/base reaction in aqueous solution, *water* is always one of the products: keeping this in mind makes predicting the formula of the *salt* produced easy to do.

    a. $HCl(aq) + RbOH(aq) \rightarrow H_2O(l) + \mathbf{RbCl}(aq)$

    b. $HClO_4(aq) + NaOH(aq) \rightarrow H_2O(l) + \mathbf{NaClO_4}(aq)$

    c. $HBr(aq) + NaOH(aq) \rightarrow H_2O(l) + \mathbf{NaBr}(aq)$

    d. $H_2SO_4(aq) + 2CsOH(aq) \rightarrow 2H_2O(l) + \mathbf{Cs_2SO_4}(aq)$

23. In general, the salt formed in an aqueous acid-base reaction consists of the *positive ion of the base* involved in the reaction, combined with the *negative ion of the acid*. The hydrogen ion of the strong acid combines with the hydroxide ion of the strong base to produce water, which is the other product of the acid-base reactions.

    a. $2NaOH(aq) + H_2SO_4(aq) \rightarrow 2H_2O(l) + Na_2SO_4(aq)$

    b. $RbOH(aq) + HNO_3(aq) \rightarrow H_2O(l) + RbNO_3(aq)$

    c. $KOH(aq) + HClO_4(aq) \rightarrow H_2O(l) + KClO_4(aq)$

    d. $KOH(aq) + HCl(aq) \rightarrow H_2O(l) + KCl(aq)$

24. A driving force, in general, is an event which tends to help to convert the reactants of a process into the products. Some elements (metals) tend to lose electrons, while other elements (nonmetals) tend to gain electrons. A *transfer* of electrons from atoms of a metal to atoms of a nonmetal would be favorable, and would result in a chemical reaction. A simple example of such a process is the reaction of sodium with chlorine: sodium atoms tend to each lose one electron (to form Na^+), whereas chlorine atoms tend to each gain one electron (to form Cl^-). The reaction of sodium metal with chlorine gas represents a transfer of electrons from sodium atoms to chlorine atoms to form sodium chloride.

25. The metallic element *loses* electrons and the nonmetallic element *gains* electrons.

26. Each potassium atom would lose one electron to become a K^+ ion. Each sulfur atom would gain two electrons to become a S^{2-} ion. Two potassium atoms would have to be oxidized to provided the two electrons needed to reduce one sulfur atom.

27. $AlBr_3$ is made up of Al^{3+} ions and Br^- ions. Aluminum atoms each lose three electrons to become Al^{3+} ions. Bromine atoms each gain one electron to become Br^- ions (so each Br_2 molecule gains two electrons to become two Br^- ions).

28. a. $K + F_2 \rightarrow KF$

    Balance fluorine: $K + F_2 \rightarrow \mathbf{2KF}$

Reactions in Aqueous Solutions

Balance potassium: $2K + F_2 \rightarrow 2KF$

Balanced equation: $2K(s) + F_2(g) \rightarrow 2KF(s)$

b. $K + O_2 \rightarrow K_2O$

Balance oxygen: $K + O_2 \rightarrow 2K_2O$

Balance potassium: $4K + O_2 \rightarrow 2K_2O$

Balanced equation: $4K(s) + O_2(g) \rightarrow 2K_2O(s)$

c. $K + N_2 \rightarrow K_3N$

Balance nitrogen: $K + N_2 \rightarrow 2K_3N$

Balance potassium: $6K + N_2 \rightarrow 2K_3N$

Balanced equation: $6K(s) + N_2(g) \rightarrow 2K_3N(s)$

d. $K + C \rightarrow K_4C$

Balance potassium: $4K + C \rightarrow K_4C$

Balanced equation: $4K(s) + C(s) \rightarrow K_4C(s)$

29. a. $Fe(s) + S(s) \rightarrow Fe_2S_3(s)$

Balance iron: $2Fe + S \rightarrow Fe_2S_3$

Balance sulfur: $2Fe + 3S \rightarrow Fe_2S_3$

Balanced equation: $2Fe(s) + 3S(s) \rightarrow Fe_2S_3(s)$

b. $Zn(s) + HNO_3(aq) \rightarrow Zn(NO_3)_2(aq) + H_2(g)$

Balance nitrate ions: $Zn + 2HNO_3 \rightarrow Zn(NO_3)_2 + H_2$

Balanced equation: $Zn(s) + 2HNO_3(aq) \rightarrow Zn(NO_3)_2(aq) + H_2(g)$

c. $Sn(s) + O_2(g) \rightarrow SnO(s)$

Balance oxygen: $Sn + O_2 \rightarrow 2SnO$

Balance tin: $2Sn + O_2 \rightarrow 2SnO$

Balanced equation: $2Sn(s) + O_2(g) \rightarrow 2SnO(s)$

d. $K(s) + H_2(g) \rightarrow KH(s)$

Balance hydrogen: $K + H_2 \rightarrow 2KH$

Balance potassium: $2K + H_2 \rightarrow 2KH$

Balanced equation: $2K(s) + H_2(g) \rightarrow 2KH(s)$

e. $Cs(s) + H_2O(l) \rightarrow CsOH(aq) + H_2(g)$

Balance hydrogen: $Cs + 2H_2O \rightarrow 2CsOH + H_2$

Balance cesium: $2Cs + 2H_2O \rightarrow 2CsOH + H_2$

Balanced equation: $2Cs(s) + 2H_2O(l) \rightarrow 2CsOH(aq) + H_2(g)$

30. A double-displacement reaction has the form AB + CD → AD + CB. In a double-displacement reaction, when two solutions of ionic solutes are mixed, the positive ions of the two solutes exchange anions (this presupposes that some driving force is present which causes a detectable reaction to occur). Two examples are:

$Pb(NO_3)_2(aq) + 2HCl(aq) \rightarrow PbCl_2(s) + 2HNO_3(aq)$

$BaCl_2(aq) + Na_2SO_4(aq) \rightarrow BaSO_4(s) + 2NaCl(aq)$

A single-displacement reaction has the form A + BC → AC + B. In a single displacement reaction, a new element replaces a less active element in its compound. Two examples are:

$Zn(s) + 2HCl(aq) \rightarrow ZnCl_2(aq) + H_2(g)$

$Cu(s) + 2AgNO_3(aq) \rightarrow Cu(NO_3)_2(aq) + 2Ag(s)$

31. examples of formation of water:

$HCl(aq) + NaOH(aq) \rightarrow H_2O(l) + NaCl(aq)$

$H_2SO_4(aq) + 2KOH(aq) \rightarrow 2H_2O(l) + K_2SO_4(aq)$

examples of formation of a gaseous product:

$Mg(s) + 2HCl(aq) \rightarrow MgCl_2(aq) + H_2(g)$

$2KClO_3(s) \rightarrow 2KCl(s) + 3O_2(g)$

32. For each reaction, the type of reaction is first identified, followed by some of the reasoning that leads to this choice (there may be more than one way in which you can recognize a particular type of reaction).

  a. precipitation ($BaSO_4$ is insoluble)

  b. oxidation-reduction (Zn changes from the elemental to the combined state; hydrogen changes from the combined to the elemental state)

  c. precipitation (AgCl is insoluble)

  d. acid-base (HCl is an acid; KOH is a base; water and a salt are produced)

  e. oxidation-reduction (Cu changes from the combined to the elemental state; Zn changes from the elemental to the combined state)

  f. acid-base (the $H_2PO_4^-$ ion behaves as an acid; NaOH behaves as a base; a salt and water are produced)

  g. precipitation ($CaSO_4$ is insoluble); acid-base [$Ca(OH)_2$ is a base; H_2SO_4 is an acid; a salt and water are produced]

  h. oxidation-reduction (Mg changes from the elemental to the combined state; Zn changes from the combined to the elemental state)

  i. precipitation ($BaSO_4$ is insoluble)

33. For each reaction, the type of reaction is first identified, followed by some of the reasoning that leads to this choice (there may be more than one way in which you can recognize a particular type of reaction).

Reactions in Aqueous Solutions   69

      a.   oxidation-reduction (oxygen changes from the combined state to the elemental state)

      b.   oxidation-reduction (copper changes from the elemental to the combined state; hydrogen changes from the combined to the elemental state)

      c.   acid-base (H_2SO_4 is a strong acid and NaOH is a strong base; water and a salt are formed)

      d.   acid-base, precipitation (H_2SO_4 is a strong acid, and $Ba(OH)_2$ is a base; water and a salt are formed; an insoluble product forms)

      e.   precipitation (AgCl is only slightly soluble)

      f.   precipitation ($Cu(OH)_2$ is only slightly soluble)

      g.   oxidation-reduction (chlorine and fluorine change from the elemental to the combined state)

      h.   oxidation-reduction (oxygen changes from the elemental to the combined state)

      i.   acid-base (HNO_3 is a strong acid and $Ca(OH)_2$ is a strong base; a salt and water are formed)

34.   A synthesis reaction represents the production of a given compound from simpler substances (either elements or simpler compounds). For example,

$$O_2(g) + 2F_2(g) \rightarrow 2OF_2(g)$$

represents a simple synthesis reaction. Synthesis reactions may often (but not necessarily always) also be classified in other ways. For example, the reaction

$$C(s) + O_2(g) \rightarrow CO_2(g)$$

could also be classified as an oxidation-reduction reaction, or as a combustion reaction (a special sub-classification of oxidation-reduction reaction that produces a flame). As another example, the reaction

$$2Fe(s) + 3Cl_2(g) \rightarrow 2FeCl_3(s)$$

is a synthesis reaction that also is an oxidation-reduction reaction.

35.   A decomposition reaction is one in which a given compound is broken down into simpler compounds or constituent elements. The reactions

$$CaCO_3(s) \rightarrow CaO(s) + CO_2(g)$$

$$2HgO(s) \rightarrow 2Hg(l) + O_2(g)$$

both represent decomposition reactions. Such reactions often (but not necessarily always) may be classified in other ways. For example, the reaction of HgO(s) is also an oxidation-reduction reaction.

36.   Compounds like those in parts b and c of this problem, containing only carbon and hydrogen, are called *hydrocarbons*. When a hydrocarbon is reacted with oxygen (O_2), the hydrocarbon is almost always converted to carbon dioxide and water vapor. Since water molecules contain an odd number of oxygen atoms, whereas O_2 contains an even number of oxygen atoms, it is often difficult to balance such equations. For this reason, it is simpler to balance the equation using fractional coefficients if necessary, and then to

multiply by a factor that will give whole number coefficients for the final balanced equation.

    a. $C_2H_5OH(l) + O_2(g) \to CO_2(g) + H_2O(g)$

        Balance carbon: $C_2H_5OH(l) + O_2(g) \to \mathbf{2}CO_2(g) + H_2O(g)$

        Balance hydrogen: $C_2H_5OH(l) + O_2(g) \to 2CO_2(g) + \mathbf{3}H_2O(g)$

        Balance oxygen: $C_2H_5OH(l) + \mathbf{3}O_2(g) \to 2CO_2(g) + 3H_2O(g)$

        Balanced equation: $C_2H_5OH(l) + 3O_2(g) \to 2CO_2(g) + 3H_2O(g)$

    b. $C_6H_{14}(l) + O_2(g) \to CO_2(g) + H_2O(g)$

        Balance carbon: $C_6H_{14}(l) + O_2(g) \to \mathbf{6}CO_2(g) + H_2O(g)$

        Balance hydrogen: $C_6H_{14}(l) + O_2(g) \to 6CO_2(g) + \mathbf{7}H_2O(g)$

        Balance oxygen: $C_6H_{14}(l) + \mathbf{(19/2)}O_2(g) \to 6CO_2(g) + 7H_2O(g)$

        Balanced equation: $2C_6H_{14}(l) + 19O_2(g) \to 12CO_2(g) + 14H_2O(g)$

    c. $C_6H_{12}(l) + O_2(g) \to CO_2(g) + H_2O(g)$

        Balance carbon: $C_6H_{12}(l) + O_2(g) \to \mathbf{6}CO_2(g) + H_2O(g)$

        Balance hydrogen: $C_6H_{12}(l) + O_2(g) \to 6CO_2(g) + \mathbf{6}H_2O(g)$

        Balanced equation: $C_6H_{12}(l) + 9O_2(g) \to 6CO_2(g) + 6H_2O(g)$

37.   a. $C_2H_6(g) + O_2(g) \to CO_2(g) + H_2O(g)$

        Balance carbon: $C_2H_6(g) + O_2(g) \to \mathbf{2}CO_2(g) + H_2O(g)$

        Balance hydrogen: $C_2H_6(g) + O_2(g) \to 2CO_2(g) + \mathbf{3}H_2O(g)$

        Balance oxygen: $C_2H_6(g) + \mathbf{(7/2)}O_2(g) \to 2CO_2(g) + 3H_2O(g)$

        Balanced equation: $2C_2H_6(g) + 7O_2(g) \to 4CO_2(g) + 6H_2O(g)$

    b. $C_2H_6O(l) + O_2(g) \to CO_2(g) + H_2O(g)$

        Balance carbon: $C_2H_6O(l) + O_2(g) \to \mathbf{2}CO_2(g) + H_2O(g)$

        Balance hydrogen: $C_2H_6O(l) + O_2(g) \to 2CO_2(g) + \mathbf{3}H_2O(g)$

        Balance oxygen: $C_2H_6O(l) + \mathbf{3}O_2(g) \to 2CO_2(g) + 3H_2O(g)$

        Balanced equation: $C_2H_6O(l) + 3O_2(g) \to 2CO_2(g) + 3H_2O(g)$

    c. $C_2H_6O_2(l) + O_2(g) \to CO_2(g) + H_2O(g)$

        Balance carbon: $C_2H_6O_2(l) + O_2(g) \to \mathbf{2}CO_2(g) + H_2O(g)$

        Balance hydrogen: $C_2H_6O_2(l) + O_2(g) \to 2CO_2(g) + \mathbf{3}H_2O(g)$

        Balance oxygen: $C_2H_6O_2(l) + \mathbf{(5/2)}O_2(g) \to 2CO_2(g) + 3H_2O(g)$

        Balanced equation: $2C_2H_6O_2(l) + 5O_2(g) \to 4CO_2(g) + 6H_2O(g)$

Reactions in Aqueous Solutions

38. a. $2Co(s) + 3S(s) \rightarrow Co_2S_3(s)$

    b. $2NO(g) + O_2(g) \rightarrow 2NO_2(g)$

    c. $FeO(s) + CO_2(g) \rightarrow FeCO_3(s)$

    d. $2Al(s) + 3F_2(g) \rightarrow 2AlF_3(s)$

    e. $2NH_3(g) + H_2CO_3(aq) \rightarrow (NH_4)_2CO_3(s)$

39. a. $2NI_3(s) \rightarrow N_2(g) + 3I_2(s)$

    b. $BaCO_3(s) \rightarrow BaO(s) + CO_2(g)$

    c. $C_6H_{12}O_6(s) \rightarrow 6C(s) + 6H_2O(g)$

    d. $Cu(NH_3)_4SO_4(s) \rightarrow CuSO_4(s) + 4NH_3(g)$

    e. $3NaN_3(s) \rightarrow Na_3N(s) + 4N_2(g)$

40. A *molecular equation* uses the normal, uncharged formulas for the compounds involved. The *complete ionic equation* shows the compounds involved broken up into their respective ions (*all* ions present are shown). The net ionic equation shows only those ions which combine to form a precipitate, a gas, or a nonionic product such as water. The *net ionic equation* shows most clearly the species that are combining with each other.

41. In several cases, the given ion may be precipitated by *many* reactants. The following are only three of the possible examples.

    a. chloride ion would precipitate when treated with solutions containing silver ion, lead(II) ion, or mercury(I) ion.

    $Ag^+(aq) + Cl^-(aq) \rightarrow AgCl(s)$

    $Pb^{2+}(aq) + 2Cl^-(aq) \rightarrow PbCl_2(s)$

    $Hg_2^{2+}(aq) + 2Cl^-(aq) \rightarrow Hg_2Cl_2(s)$

    b. calcium ion would precipitate when treated with solutions containing sulfate ion, carbonate ion, and phosphate ion.

    $Ca^{2+}(aq) + SO_4^{2-}(aq) \rightarrow CaSO_4(s)$

    $Ca^{2+}(aq) + CO_3^{2-}(aq) \rightarrow CaCO_3(s)$

    $3Ca^{2+}(aq) + 2PO_4^{3-}(aq) \rightarrow Ca_3(PO_4)_2(s)$

    c. iron(III) ion would precipitate when treated with solutions containing hydroxide, sulfide, or carbonate ions.

    $Fe^{3+}(aq) + 3OH^-(aq) \rightarrow Fe(OH)_3(s)$

    $2Fe^{3+}(aq) + 3S^{2-}(aq) \rightarrow Fe_2S_3(s)$

    $2Fe^{3+}(aq) + 3CO_3^{2-}(aq) \rightarrow Fe_2(CO_3)_3(s)$

    d. sulfate ion would precipitate when treated with solutions containing barium ion, calcium ion, or lead(II) ion.

    $Ba^{2+}(aq) + SO_4^{2-}(aq) \rightarrow BaSO_4(s)$

$Ca^{2+}(aq) + SO_4^{2-}(aq) \rightarrow CaSO_4(s)$

$Pb^{2+}(aq) + SO_4^{2-}(aq) \rightarrow PbSO_4(s)$

    e.   mercury(I) ion would precipitate when treated with solutions containing chloride ion, sulfide ion, or carbonate ion.

$Hg_2^{2+}(aq) + 2Cl^-(aq) \rightarrow Hg_2Cl_2(s)$

$Hg_2^{2+}(aq) + S^{2-}(aq) \rightarrow Hg_2S(s)$

$Hg_2^{2+}(aq) + CO_3^{2-}(aq) \rightarrow Hg_2CO_3(s)$

    f.   silver ion would precipitate when treated with solutions containing chloride ion, sulfide ion, or carbonate ion.

$Ag^+(aq) + Cl^-(aq) \rightarrow AgCl(s)$

$2Ag^+(aq) + S^{2-}(aq) \rightarrow Ag_2S(s)$

$2Ag^+(aq) + CO_3^{2-}(aq) \rightarrow Ag_2CO_3(s)$

42.   a. $2Fe^{3+}(aq) + 3CO_3^{2-}(aq) \rightarrow Fe_2(CO_3)_3(s)$

    b. $Hg_2^{2+}(aq) + 2\,Cl^-(aq) \rightarrow Hg_2Cl_2(s)$

    c.   no precipitate

    d. $Cu^{2+}(aq) + S^{2-}(aq) \rightarrow CuS(s)$

    e. $Pb^{2+}(aq) + 2Cl^-(aq) \rightarrow PbCl_2(s)$

    f. $Ca^{2+}(aq) + CO_3^{2-}(aq) \rightarrow CaCO_3(s)$

    g. $Au^{3+}(aq) + 3OH^-(aq) \rightarrow Au(OH)_3(s)$

43.   The formulas of the salts are indicated in boldface type.

    a. $HNO_3(aq) + KOH(aq) \rightarrow H_2O(l) +$ **$KNO_3(aq)$**

    b. $H_2SO_4(aq) + Ba(OH)_2(aq) \rightarrow 2H_2O(l) +$ **$BaSO_4(s)$**

    c. $HClO_4(aq) + NaOH(aq) \rightarrow H_2O(l) +$ **$NaClO_4(aq)$**

    d. $2HCl(aq) + Ca(OH)_2(aq) \rightarrow 2H_2O(l) +$ **$CaCl_2(aq)$**

44.   For each cation, the precipitates that form with the anions listed in the right-hand column are given below. If no formula is listed, it should be assumed that that anion does *not* form a precipitate with the particular cation.

Ag^+ ion: $AgCl, Ag_2CO_3, AgOH, Ag_3PO_4, Ag_2S, Ag_2SO_4$

Ba^{2+} ion: $BaCO_3, Ba(OH)_2, Ba_3(PO_4)_2, BaS, BaSO_4$

Ca^{2+} ion: $CaCO_3, Ca(OH)_2, Ca_3(PO_4)_2, CaS, CaSO_4$

Fe^{3+} ion: $Fe_2(CO_3)_3, Fe(OH)_3, FePO_4, Fe_2S_3$

Hg_2^{2+} ion: $Hg_2Cl_2, Hg_2CO_3, Hg_2(OH)_2, (Hg_2)_3(PO_4)_2, Hg_2S$

Na^+ ion: all common salts are soluble

Reactions in Aqueous Solutions

Ni^{2+} ion: $NiCO_3$, $Ni(OH)_2$, $Ni_3(PO_4)_2$, NiS

Pb^{2+} ion: $PbCl_2$, $PbCO_3$, $Pb(OH)_2$, $Pb_3(PO_4)_2$, PbS, $PbSO_4$

45. The precipitates are marked in boldface type

   a. Rule 3: AgCl is listed as an exception

   $AgNO_3(aq) + HCl(aq) \rightarrow$ **$AgCl(s)$** $+ HNO_3(aq)$

   b. Rule 6: most cabonate salts are only slightly soluble

   $CuSO_4(aq) + (NH_4)_2CO_3(aq) \rightarrow$ **$CuCO_3(s)$** $+ (NH_4)_2SO_4(aq)$

   c. Rule 6: most carbonate salts are only slightly soluble.

   $FeSO_4(aq) + K_2CO_3(aq) \rightarrow$ **$FeCO_3(s)$** $+ K_2SO_4(aq)$

   d. no reaction

   e. Rule 6: most carbonate salts are only slightly soluble

   $Pb(NO_3)_2(aq) + Li_2CO_3(aq) \rightarrow$ **$PbCO_3(s)$** $+ 2LiNO_3(aq)$

   f. Rule 5: most hydroxide compounds are only slightly soluble

   $SnCl_4(aq) + 4NaOH(aq) \rightarrow$ **$Sn(OH)_4(s)$** $+ 4NaCl(aq)$

46.
   a. Rule 3: $Ag^+(aq) + Cl^-(aq) \rightarrow AgCl(s)$
   b. Rule 6: $3Ca^{2+}(aq) + 2PO_4^{3-}(aq) \rightarrow Ca_3(PO_4)_2(s)$
   c. Rule 3: $Pb^{2+}(aq) + 2Cl^-(aq) \rightarrow PbCl_2(s)$
   d. Rule 6: $Fe^{3+}(aq) + 3OH^-(aq) \rightarrow Fe(OH)_3(s)$

47. For simplicity, the physical states of the substances are omitted.

   $2Ba + O_2 \rightarrow 2BaO$ $Ba + S \rightarrow BaS$

   $Ba + Cl_2 \rightarrow BaCl_2$ $3Ba + N_2 \rightarrow Ba_3N_2$

   $Ba + Br_2 \rightarrow BaBr_2$ $4K + O_2 \rightarrow 2K_2O$

   $2K + S \rightarrow K_2S$ $2K + Cl_2 \rightarrow 2KCl$

   $6K + N_2 \rightarrow 2K_3N$ $2K + Br_2 \rightarrow 2KBr$

   $2Mg + O_2 \rightarrow 2MgO$ $Mg + S \rightarrow MgS$

   $Mg + Cl_2 \rightarrow MgCl_2$ $3Mg + N_2 \rightarrow Mg_3N_2$

   $Mg + Br_2 \rightarrow MgBr_2$ $4Rb + O_2 \rightarrow 2Rb_2O$

   $2Rb + S \rightarrow Rb_2S$ $2Rb + Cl_2 \rightarrow 2RbCl$

   $6Rb + N_2 \rightarrow 2Rb_3N$ $2Rb + Br_2 \rightarrow 2RbBr$

   $2Ca + O_2 \rightarrow 2CaO$ $Ca + S \rightarrow CaS$

   $Ca + Cl_2 \rightarrow CaCl_2$ $3Ca + N_2 \rightarrow Ca_3N_2$

Ca + Br$_2$ → CaBr$_2$

2Li + S → Li$_2$S

6Li + N$_2$ → 2Li$_3$N

4Li + O$_2$ → 2Li$_2$O

2Li + Cl$_2$ → 2LiCl

2Li + Br$_2$ → 2LiBr

48.
a. one
b. one
c. two
d. two
e. three

49.
a. two; O + 2e$^-$ → O^{2-}
b. one; F + e$^-$ → F$^-$
c. three; N + 3e$^-$ → N^{3-}
d. one; Cl + e$^-$ → Cl$^-$
e. two; S + 2e$^-$ → S^{2-}

50.
a. 2C$_3$H$_8$O(*l*) + 9O$_2$(*g*) → 6CO$_2$(*g*) + 8H$_2$O(*g*)

oxidation-reduction, combustion

b. HCl(*aq*) + AgC$_2$H$_3$O$_2$(*aq*) → AgCl(*s*) + HC$_2$H$_3$O$_2$(*aq*)

precipitation, double-displacement

c. 3HCl(*aq*) + Al(OH)$_3$(*s*) → AlCl$_3$(*aq*) + 3H$_2$O(*l*)

acid-base, double-displacement

d. 2H$_2$O$_2$(*aq*) → 2H$_2$O(*l*) + O$_2$(*g*)

oxidation-reduction, decomposition

e. N$_2$H$_4$(*l*) + O$_2$(*g*) → N$_2$(*g*) + 2H$_2$O(*g*)

oxidation-reduction, combustion

Chapter 9 Chemical Quantities

1. Although we define mass as the "amount of matter in a substance," the *units* in which we measure mass are a human invention. Atoms and molecules react on an individual particle-by-particle basis, and we have to count individual particles when doing chemical calculations.

2. Balanced chemical equations tell us in what proportions *on a mole basis* substances combine; since the molar masses of C(s) and O_2(g) are different, 1 g of O_2 could not represent the same number of moles as 1 g of C.

3. a. $2NO(g) + O_2(g) \rightarrow 2NO_2(g)$

    Two molecules of nitrogen monoxide combine with one molecule of oxygen gas, producing two molecules of nitrogen dioxide. Two moles of gaseous nitrogen monoxide combine with one mole of gaseous oxygen, producing two moles of gaseous nitrogen dioxide.

   b. $2AgC_2H_3O_2(aq) + CuSO_4(aq) \rightarrow Ag_2SO_4(s) + Cu(C_2H_3O_2)_2(aq)$

    Note: The term "formula unit" is used in the following statement because the substances involved in the above reaction are *ionic*, and do not contain true molecules. Two formula units of silver acetate will react with one formula unit of copper(II) sulfate, precipitating one formula unit of silver sulfate and leaving one formula unit of copper(II) acetate in solution. Two moles of aqueous silver acetate react with one mole of aqueous copper(II) sulfate, to produce one mole of solid silver sulfate as a precipitate, and leaving one mole of copper(II) acetate in solution.

   c. $PCl_3(l) + 3H_2O(l) \rightarrow H_3PO_3(l) + 3HCl(g)$

    One molecule of phosphorus trichloride reacts with three molecules of water, producing one molecule of phosphorous acid and three molecules of gaseous hydrogen chloride. One mole of liquid phosphorus trichloride reacts with three moles of liquid water, producing one mole of liquid phosphorous acid and three moles of gaseous hydrogen chloride.

   d. $C_2H_6(g) + Cl_2(g) \rightarrow C_2H_5Cl(g) + HCl(g)$

    One molecule of ethane (C_2H_6) reacts with one molecule of chlorine, producing one molecule of chloroethane (C_2H_5Cl) and one molecule of hydrogen chloride. One mole of gaseous ethane combines with one mole of chlorine gas, giving one mole of gaseous chloroethane and one mole of gaseous hydrogen chloride.

4. a. $3MnO_2(s) + 4Al(s) \rightarrow 3Mn(s) + 2Al_2O_3(s)$

    Three formula units of manganese(IV) oxide react with four aluminum atoms, producing three manganese atoms and two formula units of aluminum oxide. Three moles of solid manganese(IV) oxide react with four moles of solid aluminum, to produce three moles of solid manganese and two moles of solid aluminum oxide.

   b. $B_2O_3(s) + 3CaF_2(s) \rightarrow 2BF_3(g) + 3CaO(s)$

One molecule of diboron trioxide reacts with three formula units of calcium fluoride, producing two molecules of boron trifluoride and three formula units of calcium oxide. One mole of solid diboron trioxide reacts with three moles of solid calcium fluoride, to give two moles of gaseous boron trifluoride and three moles of solid calcium oxide.

c. $3NO_2(g) + H_2O(l) \rightarrow 2HNO_3(aq) + NO(g)$

Three molecules of nitrogen dioxide [nitrogen(IV) oxide] react with one molecule of water, to produce two molecules of nitric acid and one molecule of nitrogen monoxide [nitrogen(II) oxide]. Three moles of gaseous nitrogen dioxide react with one mole of liquid water, to produce two moles of aqueous nitric acid and one mole of nitrogen monoxide gas.

d. $C_6H_6(g) + 3H_2(g) \rightarrow C_6H_{12}(g)$

One molecule of C_6H_6 (which is named benzene) reacts with three molecules of hydrogen, producing just one molecule of C_6H_{12} (which is named cyclohexane). One mole of gaseous benzene reacts with three moles of hydrogen gas, giving one mole of gaseous cyclohexane.

5. False. The coefficients of the balanced chemical equation represent the ratios on a *mole* basis by which hydrogen peroxide decomposes.

6. For converting from a given number of moles of CH_4 to the number of moles of oxygen needed for reaction, the correct mole ratio is

$$\frac{2 \text{ mol } O_2}{1 \text{ mol } CH_4}$$

For converting from a given number of moles of CH_4 to the number of moles of product produced, the ratios are

$$\frac{1 \text{ mol } CO_2}{1 \text{ mol } CH_4} \quad \text{and} \quad \frac{2 \text{ mol } H_2O}{1 \text{ mol } CH_4}$$

7. $2Ag(s) + H_2S(g) \rightarrow Ag_2S(s) + H_2(g)$

$$\frac{1 \text{ mol } Ag_2S}{2 \text{ mol } Ag} \quad \text{and} \quad \frac{1 \text{ mol } H_2}{2 \text{ mol } Ag}$$

8. a. $2FeO(s) + C(s) \rightarrow 2Fe(l) + CO_2(g)$

$$0.125 \text{ mol FeO} \times \frac{2 \text{ mol Fe}}{2 \text{ mol FeO}} = 0.125 \text{ mol Fe}$$

$$0.125 \text{ mol FeO} \times \frac{1 \text{ mol } CO_2}{2 \text{ mol FeO}} = 0.0625 \text{ mol } CO_2$$

b. $Cl_2(g) + 2KI(aq) \rightarrow 2KCl(aq) + I_2(s)$

$$0.125 \text{ mol KI} \times \frac{2 \text{ mol KCl}}{2 \text{ mol KI}} = 0.125 \text{ mol KCl}$$

Chemical Quantities

$$0.125 \text{ mol KI} \times \frac{1 \text{ mol I}_2}{2 \text{ mol KI}} = 0.0625 \text{ mol I}_2$$

    c. $Na_2B_4O_7(s) + H_2SO_4(aq) + 5H_2O(l) \rightarrow 4H_3BO_3(s) + Na_2SO_4(aq)$

$$0.125 \text{ mol Na}_2\text{B}_4\text{O}_7 \times \frac{4 \text{ mol H}_3\text{BO}_3}{1 \text{ mol Na}_2\text{B}_4\text{O}_7} = 0.500 \text{ mol H}_3\text{BO}_3$$

$$0.125 \text{ mol Na}_2\text{B}_4\text{O}_7 \times \frac{1 \text{ mol Na}_2\text{SO}_4}{1 \text{ mol Na}_2\text{B}_4\text{O}_7} = 0.125 \text{ mol Na}_2\text{SO}_4$$

    d. $CaC_2(s) + 2H_2O(l) \rightarrow Ca(OH)_2(s) + C_2H_2(g)$

$$0.125 \text{ mol CaC}_2 \times \frac{1 \text{ mol Ca(OH)}_2}{1 \text{ mol CaC}_2} = 0.125 \text{ mol Ca(OH)}_2$$

$$0.125 \text{ mol CaC}_2 \times \frac{1 \text{ mol C}_2\text{H}_2}{1 \text{ mol CaC}_2} = 0.125 \text{ mol C}_2\text{H}_2$$

9.   a. $NH_3(g) + HCl(g) \rightarrow NH_4Cl(s)$

molar mass of NH_4Cl, 53.49 g

$$0.50 \text{ mol NH}_3 \times \frac{1 \text{ mol NH}_4\text{Cl}}{1 \text{ mol NH}_3} = 0.50 \text{ mol NH}_4\text{Cl}$$

$$0.50 \text{ mol NH}_4\text{Cl} \times \frac{53.49 \text{ g NH}_4\text{Cl}}{1 \text{ mol NH}_4\text{Cl}} = 27 \text{ g NH}_4\text{Cl}$$

    b. $CH_4(g) + 4S(g) \rightarrow CS_2(l) + 2H_2S(g)$

molar masses: CS_2, 76.15 g; H_2S, 34.09 g

$$0.50 \text{ mol S} \times \frac{1 \text{ mol CS}_2}{4 \text{ mol S}} = 0.125 \text{ mol CS}_2 \,(= 0.13 \text{ mol CS}_2)$$

$$0.125 \text{ mol CS}_2 \times \frac{76.15 \text{ g CS}_2}{1 \text{ mol CS}_2} = 9.5 \text{ g CS}_2$$

$$0.50 \text{ mol S} \times \frac{2 \text{ mol H}_2\text{S}}{4 \text{ mol S}} = 0.25 \text{ mol H}_2\text{S}$$

$$0.25 \text{ mol H}_2\text{S} \times \frac{34.09 \text{ g H}_2\text{S}}{1 \text{ mol H}_2\text{S}} = 8.5 \text{ g H}_2\text{S}$$

    c. $PCl_3(l) + 3H_2O(l) \rightarrow H_3PO_3(aq) + 3HCl(aq)$

molar masses: H_3PO_3, 81.99 g; HCl, 36.46 g

$$0.50 \text{ mol PCl}_3 \times \frac{1 \text{ mol H}_3\text{PO}_3}{1 \text{ mol PCl}_3} = 0.50 \text{ mol H}_3\text{PO}_3$$

$$0.50 \text{ mol } H_3PO_3 \times \frac{81.99 \text{ g } H_3PO_3}{1 \text{ mol } H_3PO_3} = 41 \text{ g } H_3PO_3$$

$$0.50 \text{ mol } PCl_3 \times \frac{3 \text{ mol HCl}}{1 \text{ mol } PCl_3} = 1.5 \text{ mol HCl}$$

$$1.50 \text{ mol HCl} \times \frac{36.46 \text{ g HCl}}{1 \text{ mol HCl}} = 54.7 = 55 \text{ g HCl}$$

d. $NaOH(s) + CO_2(g) \rightarrow NaHCO_3(s)$

molar mass of $NaHCO_3$, 84.01 g

$$0.50 \text{ mol NaOH} \times \frac{1 \text{ mol } NaHCO_3}{1 \text{ mol NaOH}} = 0.50 \text{ mol } NaHCO_3$$

$$0.50 \text{ mol } NaHCO_3 \times \frac{84.01 \text{ g } NaHCO_3}{1 \text{ mol } NaHCO_3} = 42 \text{ g } NaHCO_3$$

10. Before doing the calculations, the equations must be *balanced*.

   a. $4KO_2(s) + 2H_2O(l) \rightarrow 3O_2(g) + 4KOH(s)$

   $$0.625 \text{ mol KOH} \times \frac{3 \text{ mol } O_2}{4 \text{ mol KOH}} = 0.469 \text{ mol } O_2$$

   b. $SeO_2(g) + 2H_2Se(g) \rightarrow 3Se(s) + 2H_2O(g)$

   $$0.625 \text{ mol } H_2O \times \frac{3 \text{ mol Se}}{2 \text{ mol } H_2O} = 0.938 \text{ mol Se}$$

   c. $2CH_3CH_2OH(l) + O_2(g) \rightarrow 2CH_3CHO(aq) + 2H_2O(l)$

   $$0.625 \text{ mol } H_2O \times \frac{2 \text{ mol } CH_3CHO}{2 \text{ mol } H_2O} = 0.625 \text{ mol } CH_3CHO$$

   d. $Fe_2O_3(s) + 2Al(s) \rightarrow 2Fe(l) + Al_2O_3(s)$

   $$0.625 \text{ mol } Al_2O_3 \times \frac{2 \text{ mol Fe}}{1 \text{ mol } Al_2O_3} = 1.25 \text{ mol Fe}$$

11. the molar mass of the substance

12. Stoichiometry is the process of using a chemical equation to calculate the relative masses of reactants and products involved in a reaction.

13. a. molar mass of Ag = 107.9 g

   $$2.01 \times 10^{-2} \text{ g Ag} \times \frac{1 \text{ mol}}{107.9 \text{ g}} = 1.86 \times 10^{-4} \text{ mol Ag}$$

   b. molar mass of $(NH_4)_2S$ = 68.15 g

Chemical Quantities

$$45.2 \text{ mg (NH}_4)_2\text{S} \times \frac{1 \text{ g}}{1000 \text{ mg}} \times \frac{1 \text{ mol}}{68.15 \text{ g}} = 6.63 \times 10^{-4} \text{ mol (NH}_4)_2\text{S}$$

c. molar mass of uranium = 238.0 g

$$61.7 \text{ μg U} \times \frac{1 \text{ g}}{10^6 \text{ μg}} \times \frac{1 \text{ mol}}{238.0 \text{ g}} = 2.59 \times 10^{-7} \text{ mol U}$$

d. molar mass of SO_2 = 64.07 g

$$5.23 \text{ kg SO}_2 \times \frac{1000 \text{ g}}{1 \text{ kg}} \times \frac{1 \text{ mol}}{64.07 \text{ g}} = 81.6 \text{ mol SO}_2$$

e. molar mass of $Fe(NO_3)_3$ = 241.9 g

$$272 \text{ g Fe(NO}_3)_3 \times \frac{1 \text{ mol}}{241.9 \text{ g}} = 1.12 \text{ mol Fe(NO}_3)_3$$

f. molar mass of $FeSO_4$ = 151.9 g

$$12.7 \text{ mg FeSO}_4 \times \frac{1 \text{ g}}{1000 \text{ mg}} \times \frac{1 \text{ mol}}{151.9 \text{ g}} = 8.36 \times 10^{-5} \text{ mol FeSO}_4$$

g. molar mass of LiOH = 23.95 g

$$6.91 \times 10^3 \text{ g LiOH} \times \frac{1 \text{ mol}}{23.95 \text{ g}} = 288.5 = 289 \text{ mol LiOH}$$

14. a. molar mass of $CaCO_3$ = 100.1 g

$$2.21 \times 10^{-4} \text{ mol CaCO}_3 \times \frac{100.1 \text{ g}}{1 \text{ mol}} = 0.0221 \text{ g CaCO}_3$$

b. molar mass of He = 4.003 g

$$2.75 \text{ mol He} \times \frac{4.003 \text{ g}}{1 \text{ mol}} = 11.0 \text{ g He}$$

c. molar mass of O_2 = 32.00 g

$$0.00975 \text{ mol O}_2 \times \frac{32.00 \text{ g}}{1 \text{ mol}} = 0.312 \text{ g O}_2$$

d. molar mass of CO_2 = 44.01 g

7.21 millimol = 0.00721 mol

$$0.00721 \text{ mol CO}_2 \times \frac{44.01 \text{ g}}{1 \text{ mol}} = 0.317 \text{ g CO}_2$$

e. molar mass of FeS = 87.92 g

$$0.835 \text{ mol FeS} \times \frac{87.92 \text{ g}}{1 \text{ mol}} = 73.4 \text{ g FeS}$$

f. molar mass of KOH = 56.11 g

$$4.01 \text{ mol KOH} \times \frac{56.11 \text{ g}}{1 \text{ mol}} = 225 \text{ g KOH}$$

g. molar mass of H_2 = 2.016 g

$$0.0219 \text{ mol } H_2 \times \frac{2.016 \text{ g}}{1 \text{ mol}} = 0.0442 \text{ g } H_2$$

15. Before any calculations are done, the equations must be *balanced*.

    a. $Mg(s) + CuCl_2(aq) \rightarrow MgCl_2(aq) + Cu(s)$

    molar mass of Mg = 24.31 g

    $$25.0 \text{ g Mg} \times \frac{1 \text{ mol}}{24.31 \text{ g}} = 1.03 \text{ mol Mg}$$

    $$1.03 \text{ mol Mg} \times \frac{1 \text{ mol CuCl}_2}{1 \text{ mol Mg}} = 1.03 \text{ mol CuCl}_2$$

    b. $2AgNO_3(aq) + NiCl_2(aq) \rightarrow 2AgCl(s) + Ni(NO_3)_2(aq)$

    molar mass of $AgNO_3$ = 169.9 g

    $$25.0 \text{ g AgNO}_3 \times \frac{1 \text{ mol}}{169.9 \text{ g}} = 0.147 \text{ mol AgNO}_3$$

    $$0.147 \text{ mol AgNO}_3 \times \frac{1 \text{ mol NiCl}_2}{2 \text{ mol AgNO}_3} = 0.0735 \text{ mol NiCl}_2$$

    c. $NaHSO_3(aq) + NaOH(aq) \rightarrow Na_2SO_3(aq) + H_2O(l)$

    molar mass of $NaHSO_3$ = 104.1 g

    $$25.0 \text{ g NaHSO}_3 \times \frac{1 \text{ mol}}{104.1 \text{ g}} = 0.240 \text{ mol NaHSO}_3$$

    $$0.240 \text{ mol NaHSO}_3 \times \frac{1 \text{ mol NaOH}}{1 \text{ mol NaHSO}_3} = 0.240 \text{ mol NaOH}$$

    d. $KHCO_3(aq) + HCl(aq) \rightarrow KCl(aq) + H_2O(l) + CO_2(g)$

    molar mass of $KHCO_3$ = 100.1 g

    $$25.0 \text{ g KHCO}_3 \times \frac{1 \text{ mol}}{100.1 \text{ g}} = 0.250 \text{ mol KHCO}_3$$

    $$0.250 \text{ mol KHCO}_3 \times \frac{1 \text{ mol HCl}}{1 \text{ mol KHCO}_3} = 0.250 \text{ mol HCl}$$

Chemical Quantities

16. Before any calculations are done, the equations must be *balanced*. Since the given and required quantities in this question are given in *milligrams*, it is most convenient to perform the calculations in terms of *millimoles* of the substances involved. One millimole of a substance represents the molar mass of the substance expressed in milligrams.

   a. $FeSO_4(aq) + K_2CO_3(aq) \rightarrow FeCO_3(s) + K_2SO_4(aq)$

   millimolar masses: $FeSO_4$, 151.9 mg; $FeCO_3$, 115.9 mg; K_2SO_4, 174.3 mg

   $$10.0 \text{ mg FeSO}_4 \times \frac{1 \text{ mmol FeSO}_4}{151.9 \text{ mg FeSO}_4} = 0.0658 \text{ mmol FeSO}_4$$

   $$0.0658 \text{ mmol FeSO}_4 \times \frac{1 \text{ mmol FeCO}_3}{1 \text{ mmol FeSO}_4} \times \frac{115.9 \text{ mg FeCO}_3}{1 \text{ mmol FeCO}_3} = 7.63 \text{ mg FeCO}_3$$

   $$0.0658 \text{ mmol FeSO}_4 \times \frac{1 \text{ mmol K}_2SO_4}{1 \text{ mmol FeSO}_4} \times \frac{174.3 \text{ mg K}_2SO_4}{1 \text{ mmol K}_2SO_4} = 11.5 \text{ mg K}_2SO_4$$

   b. $4Cr(s) + 3SnCl_4(l) \rightarrow 4CrCl_3(s) + 3Sn(s)$

   millimolar masses: Cr, 52.00 mg; $CrCl_3$, 158.4 mg; Sn, 118.7 mg

   $$10.0 \text{ mg Cr} \times \frac{1 \text{ mmol Cr}}{52.00 \text{ mg Cr}} = 0.192 \text{ mmol Cr}$$

   $$0.192 \text{ mmol Cr} \times \frac{4 \text{ mmol CrCl}_3}{4 \text{ mmol Cr}} \times \frac{158.4 \text{ mg CrCl}_3}{1 \text{ mmol CrCl}_3} = 30.4 \text{ mg CrCl}_3$$

   $$0.192 \text{ mmol Cr} \times \frac{3 \text{ mmol Sn}}{4 \text{ mmol Cr}} \times \frac{118.7 \text{ mg Sn}}{1 \text{ mmol Sn}} = 17.1 \text{ mg Sn}$$

   c. $16Fe(s) + 3S_8(s) \rightarrow 8Fe_2S_3(s)$

   millimolar masses: S_8, 256.6 mg; Fe_2S_3, 207.9 mg

   $$10.0 \text{ mg S}_8 \times \frac{1 \text{ mmol S}_8}{256.6 \text{ mg S}_8} = 0.0390 \text{ mmol S}_8$$

   $$0.0390 \text{ mmol S}_8 \times \frac{8 \text{ mmol Fe}_2S_3}{3 \text{ mmol S}_8} \times \frac{207.9 \text{ mg Fe}_2S_3}{1 \text{ mmol Fe}_2S_3} = 21.6 \text{ mg Fe}_2S_3$$

   d. $3Ag(s) + 4HNO_3(aq) \rightarrow 3AgNO_3(aq) + 2H_2O(l) + NO(g)$

   millimolar masses: HNO_3, 63.0 mg; $AgNO_3$, 169.9 mg

   H_2O, 18.0 mg; NO, 30.0 mg

   $$10.0 \text{ mg HNO}_3 \times \frac{1 \text{ mmol HNO}_3}{63.0 \text{ mg HNO}_3} = 0.159 \text{ mmol HNO}_3$$

   $$0.159 \text{ mmol HNO}_3 \times \frac{3 \text{ mmol AgNO}_3}{4 \text{ mmol HNO}_3} \times \frac{169.9 \text{ mg AgNO}_3}{1 \text{ mmol AgNO}_3} = 20.3 \text{ mg AgNO}_3$$

$$0.159 \text{ mmol HNO}_3 \times \frac{2 \text{ mmol H}_2\text{O}}{4 \text{ mmol HNO}_3} \times \frac{18.0 \text{ mg H}_2\text{O}}{1 \text{ mmol H}_2\text{O}} = 1.43 \text{ mg H}_2\text{O}$$

$$0.159 \text{ mmol HNO}_3 \times \frac{1 \text{ mmol NO}}{4 \text{ mmol HNO}_3} \times \frac{30.0 \text{ mg NO}}{1 \text{ mmol NO}} = 1.19 \text{ mg NO}$$

17. $2H_2(g) + O_2(g) \rightarrow 2H_2O(g)$

    molar masses: H_2, 2.016 g; H_2O, 18.02 g

    $$56.0 \text{ g H}_2 \times \frac{1 \text{ mol H}_2}{2.016 \text{ g H}_2} = 27.77 \text{ mol H}_2$$

    $$27.77 \text{ mol H}_2 \times \frac{2 \text{ mol H}_2\text{O}}{2 \text{ mol H}_2} = 27.77 \text{ mol H}_2\text{O}$$

    $$27.77 \text{ mol H}_2\text{O} \times \frac{18.02 \text{ g H}_2\text{O}}{1 \text{ mol H}_2\text{O}} = 500. \text{ g H}_2\text{O}$$

18. $2H_2(g) + O_2(g) \rightarrow 2H_2O(g)$

    molar masses of O_2 = 32.00 g

    $$0.0275 \text{ mol H}_2 \times \frac{1 \text{ mol O}_2}{2 \text{ mol H}_2} = 0.01375 = 0.0138 \text{ mol O}_2$$

    $$0.01375 \text{ mol O}_2 \times \frac{32.00 \text{ g O}_2}{1 \text{ mol O}_2} = 0.440 \text{ g O}_2$$

19. molar masses: C, 12.01 g; CO, 28.01 g; CO_2, 44.01 g

    $$5.00 \text{ g C} \times \frac{1 \text{ mol C}}{12.01 \text{ g C}} = 0.4163 \text{ mol C}$$

    carbon dioxide: $C(s) + O_2(g) \rightarrow CO_2(g)$

    $$0.4163 \text{ mol C} \times \frac{1 \text{ mol CO}_2}{1 \text{ mol C}} = 0.4163 \text{ mol CO}_2$$

    $$0.4163 \text{ mol CO}_2 \times \frac{44.01 \text{ g CO}_2}{1 \text{ mol CO}_2} = 18.3 \text{ g CO}_2$$

    carbon monoxide: $2C(s) + O_2(g) \rightarrow 2CO(g)$

    $$0.4163 \text{ mol C} \times \frac{2 \text{ mol CO}}{2 \text{ mol C}} = 0.4163 \text{ mol CO}$$

    $$0.4163 \text{ mol CO} \times \frac{28.01 \text{ g CO}}{1 \text{ mol CO}} = 11.7 \text{ g CO}$$

20. $2Fe(s) + 3Cl_2(g) \rightarrow 2FeCl_3(s)$

    millimolar masses: iron, 55.85 mg; $FeCl_3$, 162.2 mg

Chemical Quantities

$$15.5 \text{ mg Fe} \times \frac{1 \text{ mmol Fe}}{55.85 \text{ mg Fe}} = 0.2775 \text{ mmol Fe}$$

$$0.2775 \text{ mmol Fe} \times \frac{2 \text{ mmol FeCl}_3}{2 \text{ mmol Fe}} = 0.2775 \text{ mmol FeCl}_3$$

$$0.2775 \text{ mmol FeCl}_3 \times \frac{162.2 \text{ mg FeCl}_3}{1 \text{ mmol FeCl}_3} = 45.0 \text{ mg FeCl}_3$$

21. $2H_2O_2(aq) \rightarrow 2H_2O(l) + O_2(g)$

    molar masses: H_2O_2, 34.02 g; O_2, 32.00 g

    $$10.00 \text{ g } H_2O_2 \times \frac{1 \text{ mol } H_2O_2}{34.02 \text{ g } H_2O_2} = 0.2939 \text{ mol } H_2O_2$$

    $$0.2939 \text{ mol } H_2O_2 \times \frac{1 \text{ mol } O_2}{2 \text{ mol } H_2O_2} = 0.1470 \text{ mol } O_2$$

    $$0.1470 \text{ mol } O_2 \times \frac{32.00 \text{ g } O_2}{1 \text{ mol } O_2} = 4.704 \text{ g } O_2$$

22. $Cu(s) + S(s) \rightarrow CuS(s)$

    molar masses: Cu, 63.55 g; S, 32.07 g

    $$1.25 \text{ g Cu} \times \frac{1 \text{ mol}}{63.55 \text{ g}} = 1.97 \times 10^{-2} \text{ mol Cu}$$

    $$1.97 \times 10^{-2} \text{ mol Cu} \times \frac{1 \text{ mol S}}{1 \text{ mol Cu}} = 1.97 \times 10^{-2} \text{ mol S}$$

    $$1.97 \times 10^{-2} \text{ mol S} \times \frac{32.07 \text{ g}}{1 \text{ mol}} = 0.631 \text{ g S}$$

23. $2NH_4NO_3(s) \rightarrow 2N_2(g) + O_2(g) + 4H_2O(g)$

    molar masses: NH_4NO_3, 80.05 g; N_2, 28.02 g; O_2, 32.00 g; H_2O, 18.02 g

    $$1.25 \text{ g } NH_4NO_3 \times \frac{1 \text{ mol } NH_4NO_3}{80.05 \text{ g } NH_4NO_3} = 0.0156 \text{ mol } NH_4NO_3$$

    $$0.0156 \text{ mol } NH_4NO_3 \times \frac{2 \text{ mol } N_2}{2 \text{ mol } NH_4NO_3} = 0.0156 \text{ mol } N_2$$

    $$0.0156 \text{ mol } N_2 \times \frac{28.02 \text{ g } N_2}{1 \text{ mol } N_2} = 0.437 \text{ g } N_2$$

    $$0.0156 \text{ mol } NH_4NO_3 \times \frac{1 \text{ mol } O_2}{2 \text{ mol } NH_4NO_3} = 0.00780 \text{ mol } O_2$$

    $$0.00780 \text{ mol } O_2 \times \frac{32.00 \text{ g } O_2}{1 \text{ mol } O_2} = 0.250 \text{ g } O_2$$

$$0.0156 \text{ mol NH}_4\text{NO}_3 \times \frac{4 \text{ mol H}_2\text{O}}{2 \text{ mol NH}_4\text{NO}_3} = 0.0312 \text{ mol H}_2\text{O}$$

$$0.0312 \text{ mol H}_2\text{O} \times \frac{18.02 \text{ g H}_2\text{O}}{1 \text{ mol H}_2\text{O}} = 0.562 \text{ g H}_2\text{O}$$

As a check, note that 0.437 g + 0.250 g + 0.562 g = 1.249 g = 1.25 g.

24. $2\text{Mg}(s) + \text{O}_2(g) \rightarrow 2\text{MgO}(s)$

    molar masses: Mg, 24.31 g; MgO, 40.31 g

    $$1.25 \text{ g Mg} \times \frac{1 \text{ mol}}{24.31 \text{ g}} = 5.14 \times 10^{-2} \text{ mol Mg}$$

    $$5.14 \times 10^{-2} \text{ mol Mg} \times \frac{2 \text{ mol MgO}}{2 \text{ mol Mg}} = 5.14 \times 10^{-2} \text{ mol MgO}$$

    $$5.14 \times 10^{-2} \text{ mol MgO} \times \frac{40.31 \text{ g}}{1 \text{ mol}} = 2.07 \text{ g MgO}$$

25. From the balanced equation:

    $$\text{Cl}_2 + 2\text{KI} \rightarrow \text{I}_2 + 2\text{KCl}$$

    We can calculate the following:

    $$4.50 \times 10^3 \text{ g Cl}_2 \times \frac{1 \text{ mol Cl}_2}{70.9 \text{ g Cl}_2} \times \frac{1 \text{ mol I}_2}{1 \text{ mol Cl}_2} \times \frac{253.8 \text{ g I}_2}{1 \text{ mol I}_2} = 1.61 \times 10^4 \text{ g I}_2$$

26. From the balanced equation:

    $$\text{Cl}_2 + \text{F}_2 \rightarrow 2\text{ClF}$$

    We can calculate the following:

    $$5.00 \times 10^{-3} \text{ g ClF} \times \frac{1 \text{ mol ClF}}{54.45 \text{ g ClF}} \times \frac{1 \text{ mol Cl}_2}{2 \text{ mol ClF}} \times \frac{70.9 \text{ g Cl}_2}{1 \text{ mol Cl}_2} = 3.26 \times 10^{-3} \text{ g Cl}_2$$

27. Start by balancing the equations:

    $$\text{C}_3\text{H}_8 + 5\text{O}_2 \rightarrow 3\text{CO}_2 + 4\text{H}_2\text{O}$$

    $$2\text{C}_4\text{H}_{10} + 13\text{O}_2 \rightarrow 8\text{CO}_2 + 10\text{H}_2\text{O}$$

    If we had, for example, 10.00 g of C_3H_8 and 10.00 g of C_4H_{10}, we can calculate the mass of oxygen required for each reaction and compare the results.

    $$10.00 \text{ g C}_3\text{H}_8 \times \frac{1 \text{ mol C}_3\text{H}_8}{44.09 \text{ g C}_3\text{H}_8} \times \frac{5 \text{ mol O}_2}{1 \text{ mol C}_3\text{H}_8} \times \frac{32.00 \text{ g O}_2}{1 \text{ mol O}_2} = 36.29 \text{ g O}_2$$

    $$10.00 \text{ g C}_4\text{H}_{10} \times \frac{1 \text{ mol C}_4\text{H}_{10}}{58.12 \text{ g C}_4\text{H}_{10}} \times \frac{13 \text{ mol O}_2}{2 \text{ mol C}_4\text{H}_{10}} \times \frac{32.00 \text{ g O}_2}{1 \text{ mol O}_2} = 35.79 \text{ g O}_2$$

    Therefore, more O_2 would be required for the combustion of C_3H_8. The same result is obtained with masses other than 10.00 g.

Chemical Quantities

28. Start with the balanced equations:

$$CH_4 + 2O_2 \rightarrow CO_2 + 2H_2O$$

$$4NH_3 + 5O_2 \rightarrow 4NO + 6H_2O$$

The amount of water produced from 1.00 g CH_4 is:

$$1.00 \text{ g } CH_4 \times \frac{1 \text{ mol } CH_4}{16.042 \text{ g } CH_4} \times \frac{2 \text{ mol } H_2O}{1 \text{ mol } CH_4} = 0.125 \text{ mol } H_2O$$

The mass of NH_3 needed to produce 0.125 mol H_2O is:

$$0.125 \text{ mol } H_2O \times \frac{4 \text{ mol } NH_3}{6 \text{ mol } H_2O} \times \frac{17.034 \text{ g } NH_3}{1 \text{ mol } NH_3} = 1.42 \text{ g } NH_3$$

29. The limiting reactant is the reactant which limits the amounts of products that can form in a chemical reaction. *All* given reactants are necessary for the production of products: if the limiting reactant has been consumed, then there is none of this reactant present for reaction.

30. We start with 6 molecules N_2 and 6 molecules H_2, and these react according to the balanced equation:

$$N_2 + 3H_2 \rightarrow 2NH_3$$

We can determine amounts of product and leftover reactant from this information given.

$$N_2 + 3H_2 \rightarrow 2NH_3$$

	N_2	$3H_2$	$2NH_3$
start	6	6	0
react	-2	-6	+4
end	4	0	4

Note that these react in the same ratio as given in the balanced equation (that is, 2:6:4 = 1:3:2). The pictures should show 4 molecules N_2 and 4 molecules NH_3.

31. To determine the limiting reactant, first calculate the number of moles of each reactant present. Then determine how these numbers of moles correspond to the stoichiometric ratio indicated by the balanced chemical equation for the reaction.

32. The theoretical yield of a reaction represents the stoichiometric amount of product that should form if the limiting reactant for the process is completely consumed.

33. A reactant is present *in excess* if there is more of that reactant present than is needed to combine with the limiting reactant for the process. By definition, the limiting reactant cannot be present in excess. An excess of any reactant does not affect the theoretical yield for a process: the theoretical yield is determined by the limiting reactant.

34. a. $2Al(s) + 6HCl(aq) \rightarrow 2AlCl_3(aq) + 3H_2(g)$

Molar masses: Al, 26.98 g; HCl, 36.46 g; $AlCl_3$, 133.3 g; H_2, 2.016 g

$$15.0 \text{ g Al} \times \frac{1 \text{ mol}}{26.98 \text{ g}} = 0.556 \text{ mol Al}$$

$$15.0 \text{ g HCl} \times \frac{1 \text{ mol}}{36.46 \text{ g}} = 0.411 \text{ mol HCl}$$

Since HCl is needed to react with Al in a 6:2 (i.e., 3:1) molar ratio, it seems pretty certain that HCl is the limiting reactant. To prove this, we can calculate the quantity of Al that would react with the given number of moles of HCl.

$$0.411 \text{ mol HCl} \times \frac{2 \text{ mol Al}}{6 \text{ mol HCl}} = 0.137 \text{ mol Al}$$

By this calculation we have shown that *all* the HCl present will be needed to react with only 0.137 mol Al (out of the 0.556 mol Al present). Therefore HCl is the limiting reactant, and Al is present in excess. The calculation of the masses of products produced is based on the number of moles of the limiting reactant.

$$0.411 \text{ mol HCl} \times \frac{2 \text{ mol AlCl}_3}{6 \text{ mol HCl}} \times \frac{133.3 \text{ g}}{1 \text{ mol}} = 18.3 \text{ g AlCl}_3$$

$$0.411 \text{ mol HCl} \times \frac{3 \text{ mol H}_2}{6 \text{ mol HCl}} \times \frac{2.016 \text{ g}}{1 \text{ mol}} = 0.414 \text{ g H}_2$$

b. $2\text{NaOH}(aq) + \text{CO}_2(g) \rightarrow \text{Na}_2\text{CO}_3(aq) + \text{H}_2\text{O}(l)$

molar masses: NaOH, 40.00 g; CO_2, 44.01 g; Na_2CO_3, 105.99 g; H_2O, 18.02 g

$$15.0 \text{ g NaOH} \times \frac{1 \text{ mol}}{40.00 \text{ g}} = 0.375 \text{ mol NaOH}$$

$$15.0 \text{ g CO}_2 \times \frac{1 \text{ mol}}{44.01 \text{ g}} = 0.341 \text{ mol CO}_2$$

For the 0.375 mol NaOH, let's calculate if there is enough CO_2 present to react:

$$0.375 \text{ mol NaOH} \times \frac{1 \text{ mol CO}_2}{2 \text{ mol NaOH}} = 0.1875 \text{ mol CO}_2 \text{ (0.188 mol)}$$

We have present *more* CO_2 (0.341 mol) than is needed to react with the given quantity of NaOH. NaOH is therefore the limiting reactant, and CO_2 is present in excess. The quantities of products resulting are based on the complete conversion of the limiting reactant (0.375 mol NaOH):

$$0.375 \text{ mol NaOH} \times \frac{1 \text{ mol Na}_2\text{CO}_3}{2 \text{ mol NaOH}} \times \frac{105.99 \text{ g}}{1 \text{ mol}} = 19.9 \text{ g Na}_2\text{CO}_3$$

$$0.375 \text{ mol NaOH} \times \frac{1 \text{ mol H}_2\text{O}}{2 \text{ mol NaOH}} \times \frac{18.02 \text{ g}}{1 \text{ mol}} = 3.38 \text{ g H}_2\text{O}$$

c. $\text{Pb(NO}_3)_2(aq) + 2\text{HCl}(aq) \rightarrow \text{PbCl}_2(s) + 2\text{HNO}_3(aq)$

Molar masses: $Pb(NO_3)_2$, 331.2 g; HCl, 36.46 g; $PbCl_2$, 278.1 g; HNO_3, 63.02 g

$$15.0 \text{ g Pb(NO}_3)_2 \times \frac{1 \text{ mol}}{331.2 \text{ g}} = 0.0453 \text{ mol Pb(NO}_3)_2$$

Chemical Quantities

$$15.0 \text{ g HCl} \times \frac{1 \text{ mol}}{36.46 \text{ g}} = 0.411 \text{ mol}$$

With such a large disparity between the numbers of moles of the reactants, it's probably a sure bet that $Pb(NO_3)_2$ is the limiting reactant. To confirm this, we can calculate how many mol of HCl are needed to react with the given amount of $Pb(NO_3)_2$.

$$0.0453 \text{ mol Pb(NO}_3)_2 \times \frac{2 \text{ mol HCl}}{1 \text{ mol Pb(NO}_3)_2} = 0.0906 \text{ mol HCl}$$

We have considerably more HCl present than is needed to react completely with the $Pb(NO_3)_2$. Therefore, $Pb(NO_3)_2$ is the limiting reactant, and HCl is present in excess. The quantities of products produced are based on the limiting reactant being completely consumed.

$$0.0453 \text{ mol Pb(NO}_3)_2 \times \frac{1 \text{ mol PbCl}_2}{1 \text{ mol Pb(NO}_3)_2} \times \frac{278.1 \text{ g}}{1 \text{ mol}} = 12.6 \text{ g PbCl}_2$$

$$0.0453 \text{ mol Pb(NO}_3)_2 \times \frac{2 \text{ mol HNO}_3}{1 \text{ mol Pb(NO}_3)_2} \times \frac{63.02 \text{ g}}{1 \text{ mol}} = 5.71 \text{ g HNO}_3$$

d. $2K(s) + I_2(s) \rightarrow 2KI(s)$

Molar masses: K, 39.10 g; I_2, 253.8 g; KI, 166.0 g

$$15.0 \text{ g K} \times \frac{1 \text{ mol}}{39.10 \text{ g}} = 0.384 \text{ mol K}$$

$$15.0 \text{ g I}_2 \times \frac{1 \text{ mol}}{253.8 \text{ g}} = 0.0591 \text{ mol I}_2$$

Since, from the balanced chemical equation, we need twice as many moles of K as moles of I_2, and since there is so little I_2 present, it's a safe bet that I_2 is the limiting reactant. To confirm this, we can calculate how many moles of K are needed to react with the given amount of I_2 present:

$$0.0591 \text{ mol I}_2 \times \frac{2 \text{ mol K}}{1 \text{ mol I}_2} = 0.1182 \text{ mol K}$$

Clearly we have more potassium present than is needed to react with the small amount of I_2 present. I_2 is therefore the limiting reactant, and potassium is present in excess. The amount of product produced is calculated from the number of moles of the limiting reactant present:

$$0.0591 \text{ mol I}_2 \times \frac{2 \text{ mol KI}}{1 \text{ mol I}_2} \times \frac{166.0 \text{ g}}{1 \text{ mol}} = 19.6 \text{ g KI}$$

35. a. $2NH_3(g) + 2Na(s) \rightarrow 2NaNH_2(s) + H_2(g)$

Molar masses: NH_3, 17.03 g; Na, 22.99 g; $NaNH_2$, 39.02 g

$$50.0 \text{ g NH}_3 \times \frac{1 \text{ mol}}{17.03 \text{ g}} = 2.94 \text{ mol NH}_3$$

$$50.0 \text{ g Na} \times \frac{1 \text{ mol}}{22.99 \text{ g}} = 2.17 \text{ mol Na}$$

Since the coefficients of NH_3 and Na are the *same* in the balanced chemical equation for the reaction, the two reactants combine in a 1:1 molar ratio. Therefore Na is the limiting reactant, which will control the amount of product produced.

$$2.17 \text{ mol Na} \times \frac{2 \text{ mol NaNH}_2}{2 \text{ mol Na}} \times \frac{39.02 \text{ g}}{1 \text{ mol}} = 84.7 \text{ g NaNH}_2$$

b. $BaCl_2(aq) + Na_2SO_4(aq) \rightarrow BaSO_4(s) + 2NaCl(aq)$

Molar masses: $BaCl_2$, 208.2 g; Na_2SO_4, 142.1 g; $BaSO_4$, 233.4 g

$$50.0 \text{ g BaCl}_2 \times \frac{1 \text{ mol}}{208.2 \text{ g}} = 0.240 \text{ mol BaCl}_2$$

$$50.0 \text{ g Na}_2SO_4 \times \frac{1 \text{ mol}}{142.1 \text{ g}} = 0.352 \text{ mol Na}_2SO_4$$

Since the coefficients of $BaCl_2$ and Na_2SO_4 are the same in the balanced chemical equation for the reaction, the reactant having the smaller number of moles present ($BaCl_2$) must be the limiting reactant, which will control the amount of product produced.

$$0.240 \text{ mol BaCl}_2 \times \frac{1 \text{ mol BaSO}_4}{1 \text{ mol BaCl}_2} \times \frac{233.4 \text{ g}}{1 \text{ mol}} = 56.0 \text{ g BaSO}_4$$

c. $SO_2(g) + 2NaOH(aq) \rightarrow Na_2SO_3(aq) + H_2O(l)$

Molar masses: SO_2, 64.07 g; NaOH, 40.00 g; Na_2SO_3, 126.1 g

$$50.0 \text{ g SO}_2 \times \frac{1 \text{ mol}}{64.07 \text{ g}} = 0.780 \text{ mol SO}_2$$

$$50.0 \text{ g NaOH} \times \frac{1 \text{ mol}}{40.00 \text{ g}} = 1.25 \text{ mol NaOH}$$

From the balanced chemical equation for the reaction, every time one mol of SO_2 reacts, two mol of NaOH are needed. For 0.780 mol of SO_2, 2(0.780 mol) = 1.56 mol of NaOH would be needed. We do not have sufficient NaOH to react with the SO_2 present: therefore, NaOH is the limiting reactant, and controls the amount of product obtained.

$$1.25 \text{ mol NaOH} \times \frac{1 \text{ mol Na}_2SO_3}{2 \text{ mol NaOH}} \times \frac{126.1 \text{ g}}{1 \text{ mol}} = 78.8 \text{ g Na}_2SO_3$$

d. $2Al(s) + 3H_2SO_4(l) \rightarrow Al_2(SO_4)_3(s) + 3H_2(g)$

Molar masses: Al, 26.98 g; H_2SO_4, 98.09 g; $Al_2(SO_4)_3$, 342.2 g

$$50.0 \text{ g Al} \times \frac{1 \text{ mol}}{26.98 \text{ g}} = 1.85 \text{ mol Al}$$

$$50.0 \text{ g H}_2SO_4 \times \frac{1 \text{ mol}}{98.09 \text{ g}} = 0.510 \text{ mol H}_2SO_4$$

Chemical Quantities

Since the amount of H_2SO_4 present is smaller that the amount of Al, let's see if H_2SO_4 is the limiting reactant, be calculating how much Al would react with the given amount of H_2SO_4.

$$0.510 \text{ mol } H_2SO_4 \times \frac{2 \text{ mol Al}}{3 \text{ mol } H_2SO_4} = 0.340 \text{ mol Al}$$

Since all the H_2SO_4 present would react with only a small portion of the Al present, H_2SO_4 is therefore the limiting reactant and will control the amount of product obtained.

$$0.510 \text{ mol } H_2SO_4 \times \frac{1 \text{ mol } Al_2(SO_4)_3}{3 \text{ mol } H_2SO_4} \times \frac{342.2 \text{ g}}{1 \text{ mol}} = 58.2 \text{ g } Al_2(SO_4)_3$$

36. a. $CO(g) + 2H_2(g) \rightarrow CH_3OH(l)$

    CO is the limiting reactant; 11.4 mg CH_3OH

  b. $2Al(s) + 3I_2(s) \rightarrow 2AlI_3(s)$

    I_2 is the limiting reactant; 10.7 mg AlI_3

  c. $Ca(OH)_2(aq) + 2HBr(aq) \rightarrow CaBr_2(aq) + 2H_2O(l)$

    HBr is the limiting reactant; 12.4 mg $CaBr_2$; 2.23 mg H_2O

  d. $2Cr(s) + 2H_3PO_4(aq) \rightarrow 2CrPO_4(s) + 3H_2(g)$

    H_3PO_4 is the limiting reactant; 15.0 mg $CrPO_4$; 0.309 mg H_2

37. $Cl_2(g) + 2NaI(aq) \rightarrow 2NaCl(aq) + I_2(s)$

    $Br_2(l) + 2NaI(aq) \rightarrow 2NaBr(aq) + I_2(s)$

    molar masses: Cl_2, 70.90 g; Br_2, 159.8 g; I_2, 253.8 g; NaI, 149.9 g

    $$5.00 \text{ g } Cl_2 \times \frac{1 \text{ mol } Cl_2}{70.90 \text{ g } Cl_2} = 0.0705 \text{ mol } Cl_2$$

    $$25.0 \text{ g NaI} \times \frac{1 \text{ mol NaI}}{149.9 \text{ g NaI}} = 0.167 \text{ mol NaI}$$

    Since we would need 2(0.0705) = 0.141 mol of NaI for the Cl_2 to react completely, and we have more than this amount of NaI, then Cl_2 must be the limiting reactant.

    $$0.0705 \text{ mol } Cl_2 \times \frac{1 \text{ mol } I_2}{1 \text{ mol } Cl_2} = 0.0705 \text{ mol } I_2$$

    $$0.0705 \text{ mol } I_2 \times \frac{253.8 \text{ g } I_2}{1 \text{ mol } I_2} = 17.9 \text{ g } I_2$$

    $$5.00 \text{ g } Br_2 \times \frac{1 \text{ mol } Br_2}{159.8 \text{ g } Br_2} = 0.0313 \text{ mol } Br_2$$

$$25.0 \text{ g NaI} \times \frac{1 \text{ mol NaI}}{149.9 \text{ g NaI}} = 0.167 \text{ mol NaI}$$

Since we would need 2(0.0313) = 0.0626 mol of NaI for the Br_2 to react completely, and we have more than this amount of NaI, then Br_2 must be the limiting reactant.

$$0.0313 \text{ mol Br}_2 \times \frac{1 \text{ mol I}_2}{1 \text{ mol Br}_2} = 0.0313 \text{ mol I}_2$$

$$0.0313 \text{ mol I}_2 \times \frac{253.8 \text{ g I}_2}{1 \text{ mol I}_2} = 7.94 \text{ g I}_2$$

38. $4Fe(s) + 3O_2(g) \rightarrow 2Fe_2O_3(s)$

    Molar masses: Fe, 55.85 g; Fe_2O_3, 159.7 g

    $$1.25 \text{ g Fe} \times \frac{1 \text{ mol}}{55.85 \text{ g}} = 0.0224 \text{ mol Fe present}$$

    Calculate how many mol of O_2 are required to react with this amount of Fe.

    $$0.0224 \text{ mol Fe} \times \frac{3 \text{ mol O}_2}{4 \text{ mol Fe}} = 0.0168 \text{ mol O}_2$$

    Since we have more O_2 than this, Fe must be the limiting reactant.

    $$0.0224 \text{ mol Fe} \times \frac{2 \text{ mol Fe}_2O_3}{4 \text{ mol Fe}} = 0.0112 \text{ mol Fe}_2O_3$$

    $$0.0112 \text{ mol Fe}_2O_3 \times \frac{159.7 \text{ g Fe}_2O_3}{1 \text{ mol Fe}_2O_3} = 1.79 \text{ g Fe}_2O_3$$

39. $Ca^{2+}(aq) + Na_2C_2O_4(aq) \rightarrow CaC_2O_4(s) + 2Na^+(aq)$

    molar masses: Ca^{2+}, 40.08 g; $Na_2C_2O_4$, 134.0 g

    $$15 \text{ g Ca}^{2+} \times \frac{1 \text{ mol Ca}^{2+}}{40.08 \text{ g Ca}^{2+}} = 0.37 \text{ mol Ca}^{2+}$$

    $$15 \text{ g Na}_2C_2O_4 \times \frac{1 \text{ mol Na}_2C_2O_4}{134.0 \text{ g Na}_2C_2O_4} = 0.11 \text{ mol Na}_2C_2O_4$$

    Since the balanced chemical equation tells us that one oxalate ion is needed to precipitate each calcium ion, from the number of moles calculated to be present, it should be clear that not nearly enough sodium oxalate ion has been added to precipitate all the calcium ion in the sample.

40. The actual yield for a reaction is the quantity of product *actually isolated* from the reaction vessel. The theoretical yield represents the mass of products that should be produced by the reaction if the limiting reactant is fully consumed. The percent yield represents what *fraction* of the amount of product that *should* have been collected was *actually* collected.

Chemical Quantities

41. If the reaction is performed in a solvent, the product may have a substantial solubility in the solvent; the reaction may come to equilibrium before the full yield of product is achieved; loss of product may occur through operator error.

42. Percent yield = $\dfrac{\text{actual yield}}{\text{theoretical yield}} \times 100 = \dfrac{1.23 \text{ g}}{1.44 \text{ g}} \times 100 = 85.4\%$

43. $S_8(s) + 8Na_2SO_3(aq) + 40H_2O(l) \rightarrow 8Na_2S_2O_3 \cdot 5H_2O$

    molar masses: S_8, 256.6 g; Na_2SO_3, 126.1 g; $Na_2S_2O_3 \cdot 5H_2O$, 248.2 g

    $3.25 \text{ g } S_8 \times \dfrac{1 \text{ mol } S_8}{256.6 \text{ g } S_8} = 0.01267 \text{ mol } S_8$

    $13.1 \text{ g } Na_2SO_3 \times \dfrac{1 \text{ mol } Na_2SO_3}{126.1 \text{ g } Na_2SO_3} = 0.1039 \text{ mol } Na_2SO_3$

    S_8 is the limiting reactant.

    $0.01267 \text{ mol } S_8 \times \dfrac{8 \text{ mol } Na_2S_2O_3 \cdot 5H_2O}{1 \text{ mol } S_8} = 0.1014 \text{ mol } Na_2S_2O_3 \cdot 5H_2O$

    $0.1014 \text{ mol } Na_2S_2O_3 \cdot 5H_2O \times \dfrac{248.2 \text{ g } Na_2S_2O_3 \cdot 5H_2O}{1 \text{ mol } Na_2S_2O_3 \cdot 5H_2O} = 25.2 \text{ g } Na_2S_2O_3 \cdot 5H_2O$

    Percent yield = $\dfrac{\text{actual yield}}{\text{theoretical yield}} \times 100 = \dfrac{5.26 \text{ g}}{25.2 \text{ g}} \times 100 = 20.9\%$

44. $2LiOH(s) + CO_2(g) \rightarrow Li_2CO_3(s) + H_2O(g)$

    molar masses: LiOH, 23.95 g; CO_2, 44.01 g

    $155 \text{ g LiOH} \times \dfrac{1 \text{ mol LiOH}}{23.95 \text{ g LiOH}} \times \dfrac{1 \text{ mol } CO_2}{2 \text{ mol LiOH}} \times \dfrac{44.01 \text{ g } CO_2}{1 \text{ mol } CO_2} = 142 \text{ g } CO_2$

    Since the cartridge has only absorbed 102 g CO_2 out of a total capacity of 142 g CO_2, the cartridge has absorbed

    $\dfrac{102 \text{ g}}{142 \text{ g}} \times 100 = 71.8\%$ of its capacity

45. $Xe(g) + 2F_2(g) \rightarrow XeF_4(s)$

    molar masses: Xe, 131.3 g; F_2, 38.00 g; XeF_4, 207.3 g

    $130. \text{ g Xe} \times \dfrac{1 \text{ mol Xe}}{131.3 \text{ g Xe}} = 0.9901 \text{ mol Xe}$

    $100. \text{ g } F_2 \times \dfrac{1 \text{ mol } F_2}{38.00 \text{ g } F_2} = 2.632 \text{ mol } F_2$

    Xe is the limiting reactant.

$$0.9901 \text{ mol Xe} \times \frac{1 \text{ mol XeF}_4}{1 \text{ mol Xe}} = 0.9901 \text{ mol XeF}_4$$

$$0.9901 \text{ mol XeF}_4 \times \frac{207.3 \text{ g XeF}_4}{1 \text{ mol XeF}_4} = 205 \text{ g XeF}_4$$

Percent yield = $\dfrac{\text{actual yield}}{\text{theoretical yield}} \times 100 = \dfrac{145 \text{ g}}{205 \text{ g}} \times 100 = 70.7\%$

46. $CaCO_3(s) + 2HCl(g) \rightarrow CaCl_2(s) + CO_2(g) + H_2O(g)$

    molar masses: $CaCO_3$, 100.1 g; HCl, 36.46 g; $CaCl_2$, 111.0 g

    $$155 \text{ g CaCO}_3 \times \frac{1 \text{ mol CaCO}_3}{100.1 \text{ g CaCO}_3} = 1.548 \text{ mol CaCO}_3$$

    $$250. \text{ g HCl} \times \frac{1 \text{ mol HCl}}{36.46 \text{ g HCl}} = 6.857 \text{ mol HCl}$$

    $CaCO_3$ is the limiting reactant.

    $$1.548 \text{ mol CaCO}_3 \times \frac{1 \text{ mol CaCl}_2}{1 \text{ mol CaCO}_3} = 1.548 \text{ mol CaCl}_2$$

    $$1.548 \text{ mol CaCl}_2 \times \frac{111.0 \text{ g CaCl}_2}{1 \text{ mol CaCl}_2} = 172 \text{ g CaCl}_2$$

    Percent yield = $\dfrac{\text{actual yield}}{\text{theoretical yield}} \times 100 = \dfrac{142 \text{ g}}{172 \text{ g}} \times 100 = 82.6\%$

47. $NaCl(aq) + NH_3(aq) + H_2O(l) + CO_2(s) \rightarrow NH_4Cl(aq) + NaHCO_3(s)$

    molar masses: NH_3, 17.03 g; CO_2, 44.01 g; $NaHCO_3$, 84.01 g

    $$10.0 \text{ g NH}_3 \times \frac{1 \text{ mol NH}_3}{17.03 \text{ g NH}_3} = 0.5872 \text{ mol NH}_3$$

    $$15.0 \text{ g CO}_2 \times \frac{1 \text{ mol CO}_2}{44.01 \text{ g CO}_2} = 0.3408 \text{ mol CO}_2$$

    CO_2 is the limiting reactant.

    $$0.3408 \text{ mol CO}_2 \times \frac{1 \text{ mol NaHCO}_3}{1 \text{ mol CO}_2} = 0.3408 \text{ mol NaHCO}_3$$

    $$0.3408 \text{ mol NaHCO}_3 \times \frac{84.01 \text{ g NaHCO}_3}{1 \text{ mol NaHCO}_3} = 28.6 \text{ g NaHCO}_3$$

48. $Fe(s) + S(s) \rightarrow FeS(s)$

    molar masses: Fe, 55.85 g; S, 32.07 g; FeS, 87.92 g

    $$5.25 \text{ g Fe} \times \frac{1 \text{ mol Fe}}{55.85 \text{ g Fe}} = 0.0940 \text{ mol Fe}$$

Chemical Quantities

$$12.7 \text{ g S} \times \frac{1 \text{ mol S}}{32.07 \text{ g S}} = 0.396 \text{ mol}$$

Fe is the limiting reactant.

$$0.0940 \text{ mol Fe} \times \frac{1 \text{ mol FeS}}{1 \text{ mol Fe}} \times \frac{87.92 \text{ g FeS}}{1 \text{ mol FeS}} = 8.26 \text{ g FeS produced}$$

49. $C_6H_{12}O_6(s) + 6O_2(g) \rightarrow 6CO_2(g) + 6H_2O(g)$

    molar masses: glucose, 180.2 g; CO_2, 44.01 g

    $$1.00 \text{ g glucose} \times \frac{1 \text{ mol glucose}}{180.2 \text{ g glucose}} = 5.549 \times 10^{-3} \text{ mol glucose}$$

    $$5.549 \times 10^{-3} \text{ mol glucose} \times \frac{6 \text{ mol CO}_2}{1 \text{ mol glucose}} = 3.33 \times 10^{-2} \text{ mol CO}_2$$

    $$3.33 \times 10^{-2} \text{ mol CO}_2 \times \frac{44.01 \text{ g CO}_2}{1 \text{ mol CO}_2} = 1.47 \text{ g CO}_2$$

50. mass of Cl^- present $= 1.054 \text{ g sample} \times \dfrac{10.3 \text{ g Cl}^-}{100.0 \text{ g sample}} = 0.1086 \text{ g Cl}^-$

    molar masses: Cl^-, 35.45 g; $AgNO_3$, 169.9 g; $AgCl$, 143.4 g

    $$0.1086 \text{ g Cl}^- \times \frac{1 \text{ mol Cl}^-}{35.45 \text{ g Cl}^-} = 3.063 \times 10^{-3} \text{ mol Cl}^-$$

    $$3.063 \times 10^{-3} \text{ mol Cl}^- \times \frac{1 \text{ mol AgNO}_3}{1 \text{ mol Cl}^-} = 3.063 \times 10^{-3} \text{ mol AgNO}_3$$

    $$3.063 \times 10^{-3} \text{ mol AgNO}_3 \times \frac{169.9 \text{ g AgNO}_3}{1 \text{ mol AgNO}_3} = 0.520 \text{ g AgNO}_3 \text{ required}$$

    $$3.063 \times 10^{-3} \text{ mol Cl}^- \times \frac{1 \text{ mol AgCl}}{1 \text{ mol Cl}^-} = 3.063 \times 10^{-3} \text{ mol AgCl}$$

    $$3.063 \times 10^{-3} \text{ mol AgCl} \times \frac{143.4 \text{ g AgCl}}{1 \text{ mol AgCl}} = 0.439 \text{ g AgCl produced}$$

51. For O_2: $\dfrac{5 \text{ mol O}_2}{1 \text{ mol C}_3H_8}$ For CO_2: $\dfrac{3 \text{ mol CO}_2}{1 \text{ mol C}_3H_8}$ For H_2O: $\dfrac{4 \text{ mol H}_2O}{1 \text{ mol C}_3H_8}$

52. a. $2H_2O_2(l) \rightarrow 2H_2O(l) + O_2(g)$

    $$0.50 \text{ mol H}_2O_2 \times \frac{2 \text{ mol H}_2O}{2 \text{ mol H}_2O_2} = 0.50 \text{ mol H}_2O$$

    $$0.50 \text{ mol H}_2O_2 \times \frac{1 \text{ mol O}_2}{2 \text{ mol H}_2O_2} = 0.25 \text{ mol O}_2$$

    b. $2KClO_3(s) \rightarrow 2KCl(s) + 3O_2(g)$

$$0.50 \text{ mol KClO}_3 \times \frac{2 \text{ mol KCl}}{2 \text{ mol KClO}_3} = 0.50 \text{ mol KCl}$$

$$0.50 \text{ mol KClO}_3 \times \frac{3 \text{ mol O}_2}{2 \text{ mol KClO}_3} = 0.75 \text{ mol O}_2$$

c. $2Al(s) + 6HCl(aq) \rightarrow 2AlCl_3(aq) + 3H_2(g)$

$$0.50 \text{ mol Al} \times \frac{2 \text{ mol AlCl}_3}{2 \text{ mol Al}} = 0.50 \text{ mol AlCl}_3$$

$$0.50 \text{ mol Al} \times \frac{3 \text{ mol H}_2}{2 \text{ mol Al}} = 0.75 \text{ mol H}_2$$

d. $C_3H_8(g) + 5O_2(g) \rightarrow 3CO_2(g) + 4H_2O(l)$

$$0.50 \text{ mol C}_3\text{H}_8 \times \frac{3 \text{ mol CO}_2}{1 \text{ mol C}_3\text{H}_8} = 1.5 \text{ mol CO}_2$$

$$0.50 \text{ mol C}_3\text{H}_8 \times \frac{4 \text{ mol H}_2\text{O}}{1 \text{ mol C}_3\text{H}_8} = 2.0 \text{ mol H}_2\text{O}$$

53. a. $NH_3(g) + HCl(g) \rightarrow NH_4Cl(s)$

molar mass of $NH_3 = 17.0$ g

$$1.00 \text{ g NH}_3 \times \frac{1 \text{ mol NH}_3}{17.0 \text{ g NH}_3} = 0.0588 \text{ mol NH}_3$$

$$0.0588 \text{ mol NH}_3 \times \frac{1 \text{ mol NH}_4\text{Cl}}{1 \text{ mol NH}_3} = 0.0588 \text{ mol NH}_4\text{Cl}$$

b. $CaO(s) + CO_2(g) \rightarrow CaCO_3(s)$

molar mass $CaO = 56.1$ g

$$1.00 \text{ g CaO} \times \frac{1 \text{ mol CaO}}{56.1 \text{ g CaO}} = 0.0178 \text{ mol CaO}$$

$$0.0178 \text{ mol CaO} \times \frac{1 \text{ mol CaCO}_3}{1 \text{ mol CaO}} = 0.0178 \text{ mol CaCO}_3$$

c. $4Na(s) + O_2(g) \rightarrow 2Na_2O(s)$

molar mass $Na = 22.99$ g

$$1.00 \text{ g Na} \times \frac{1 \text{ mol Na}}{22.99 \text{ g Na}} = 0.0435 \text{ mol Na}$$

$$0.0435 \text{ mol Na} \times \frac{2 \text{ mol Na}_2\text{O}}{4 \text{ mol Na}} = 0.0217 \text{ mol Na}_2\text{O}$$

d. $2P(s) + 3Cl_2(g) \rightarrow 2PCl_3(l)$

Chemical Quantities

molar mass P = 30.97 g

$$1.00 \text{ g P} \times \frac{1 \text{ mol P}}{30.97 \text{ g P}} = 0.0323 \text{ mol P}$$

$$0.0323 \text{ mol P} \times \frac{2 \text{ mol PCl}_3}{2 \text{ mol P}} = 0.0323 \text{ mol PCl}_3$$

54. $2Na_2O_2(s) + 2H_2O(l) \rightarrow 4NaOH(aq) + O_2(g)$

    molar masses: Na_2O_2, 77.98 g; O_2, 32.00 g

    $$3.25 \text{ g Na}_2O_2 \times \frac{1 \text{ mol Na}_2O_2}{77.98 \text{ g Na}_2O_2} = 0.0417 \text{ mol Na}_2O_2$$

    $$0.0417 \text{ mol Na}_2O_2 \times \frac{1 \text{ mol O}_2}{2 \text{ mol Na}_2O_2} = 0.0209 \text{ mol O}_2$$

    $$0.0209 \text{ mol O}_2 \times \frac{32.00 \text{ g O}_2}{1 \text{ mol O}_2} = 0.669 \text{ g O}_2$$

55. $2C_2H_2(g) + 5O_2(g) \rightarrow 4CO_2(g) + 2H_2O(g)$

    molar masses: C_2H_2, 26.04 g; O_2, 32.00 g   150 g = 1.5 x 10^2 g

    $$1.5 \times 10^2 \text{ g C}_2H_2 \times \frac{1 \text{ mol C}_2H_2}{26.04 \text{ g C}_2H_2} = 5.760 \text{ mol C}_2H_2$$

    $$5.760 \text{ mol C}_2H_2 \times \frac{5 \text{ mol O}_2}{2 \text{ mol C}_2H_2} = 14.40 \text{ mol O}_2$$

    $$14.40 \text{ mol O}_2 \times \frac{32.00 \text{ g O}_2}{1 \text{ mol O}_2} = 4.6 \times 10^2 \text{ g O}_2$$

56. a. $C_2H_5OH(l) + 3O_2(g) \rightarrow 2CO_2(g) + 3H_2O(l)$

    molar masses: C_2H_5OH, 46.07 g; O_2, 32.00 g; CO_2, 44.01 g

    $$25.0 \text{ g C}_2H_5OH \times \frac{1 \text{ mol C}_2H_5OH}{46.07 \text{ g C}_2H_5OH} = 0.5427 \text{ mol C}_2H_5OH$$

    $$25.0 \text{ g O}_2 \times \frac{1 \text{ mol O}_2}{32.00 \text{ g O}_2} = 0.7813 \text{ mol O}_2$$

    Since there is less C_2H_5OH present on a mole basis, see if this substance is the limiting reactant.

    $$0.5427 \text{ mol C}_2H_5OH \times \frac{3 \text{ mol O}_2}{1 \text{ mol C}_2H_5OH} = 1.6281 \text{ mol O}_2$$

    From the above calculation, C_2H_5OH must *not* be the limiting reactant (even though there is a smaller number of moles of C_2H_5OH present) since more oxygen than is present would be required to react completely with the C_2H_5OH present. Oxygen is the limiting reactant.

$$0.7813 \text{ mol } O_2 \times \frac{2 \text{ mol } CO_2}{3 \text{ mol } O_2} = 0.5209 \text{ mol } CO_2$$

$$0.5209 \text{ mol } CO_2 \times \frac{44.01 \text{ g } CO_2}{1 \text{ mol } CO_2} = 22.9 \text{ g } CO_2$$

b. $N_2(g) + O_2(g) \rightarrow 2NO(g)$

molar masses: N_2, 28.02 g; O_2, 32.00 g; NO, 30.01 g

$$25.0 \text{ g } N_2 \times \frac{1 \text{ mol } N_2}{28.02 \text{ g } N_2} = 0.8922 \text{ mol } N_2$$

$$25.0 \text{ g } O_2 \times \frac{1 \text{ mol } O_2}{32.00 \text{ g } O_2} = 0.7813 \text{ mol } O_2$$

Since the coefficients of N_2 and O_2 are the *same* in the balanced chemical equation for the reaction, an equal number of moles of each substance would be necessary for complete reaction. Since there is less O_2 present on a mole basis, O_2 must be the limiting reactant.

$$0.7813 \text{ mol } O_2 \times \frac{2 \text{ mol NO}}{1 \text{ mol } O_2} = 1.5626 \text{ mol NO}$$

$$1.5626 \text{ mol NO} \times \frac{30.01 \text{ g NO}}{1 \text{ mol NO}} = 46.9 \text{ g NO}$$

c. $2NaClO_2(aq) + Cl_2(g) \rightarrow 2ClO_2(g) + 2NaCl(aq)$

molar masses: $NaClO_2$, 90.44 g; Cl_2, 70.90 g; NaCl, 58.44 g

$$25.0 \text{ g } NaClO_2 \times \frac{1 \text{ mol } NaClO_2}{90.44 \text{ g } NaClO_2} = 0.2764 \text{ mol } NaClO_2$$

$$25.0 \text{ g } Cl_2 \times \frac{1 \text{ mol } Cl_2}{70.90 \text{ g } Cl_2} = 0.3526 \text{ mol } Cl_2$$

See if $NaClO_2$ is the limiting reactant.

$$0.2764 \text{ mol } NaClO_2 \times \frac{1 \text{ mol } Cl_2}{1 \text{ mol } NaClO_2} = 0.1382 \text{ mol } Cl_2$$

Since 0.2764 mol of $NaClO_2$ would require only 0.1382 mol Cl_2 to react completely (and since we have more than this amount of Cl_2), then $NaClO_2$ must indeed be the limiting reactant.

$$0.2764 \text{ mol } NaClO_2 \times \frac{2 \text{ mol NaCl}}{2 \text{ mol } NaClO_2} = 0.2764 \text{ mol NaCl}$$

$$0.2764 \text{ mol NaCl} \times \frac{58.44 \text{ g NaCl}}{1 \text{ mol NaCl}} = 16.2 \text{ g NaCl}$$

d. $3H_2(g) + N_2(g) \rightarrow 2NH_3(g)$

Chemical Quantities

molar masses: H_2, 2.016 g; N_2, 28.02 g; NH_3, 17.03 g

$$25.0 \text{ g } H_2 \times \frac{1 \text{ mol } H_2}{2.016 \text{ g } H_2} = 12.40 \text{ mol } H_2$$

$$25.0 \text{ g } N_2 \times \frac{1 \text{ mol } N_2}{28.02 \text{ g } N_2} = 0.8922 \text{ mol } N_2$$

See if N_2 is the limiting reactant.

$$0.8922 \text{ mol } N_2 \times \frac{3 \text{ mol } H_2}{1 \text{ mol } N_2} = 2.677 \text{ mol } H_2$$

N_2 is clearly the limiting reactant, since there is 12.40 mol H_2 present (a large excess).

$$0.8922 \text{ mol } N_2 \times \frac{2 \text{ mol } NH_3}{1 \text{ mol } N_2} = 1.784 \text{ mol } NH_3$$

$$1.784 \text{ mol } NH_3 \times \frac{17.03 \text{ g } NH_3}{1 \text{ mol } NH_3} = 30.4 \text{ g } NH_3$$

57. $N_2H_4(l) + O_2(g) \rightarrow N_2(g) + 2H_2O(g)$

molar masses: N_2H_4, 32.05 g; O_2, 32.00 g; N_2, 28.02 g; H_2O, 18.02 g

$$20.0 \text{ g } N_2H_4 \times \frac{1 \text{ mol } N_2H_2}{32.05 \text{ g } N_2H_2} = 0.624 \text{ mol } N_2H_4$$

$$20.0 \text{ g } O_2 \times \frac{1 \text{ mol } O_2}{32.00 \text{ g } O_2} = 0.625 \text{ mol } O_2$$

The two reactants are present in nearly the required ratio for complete reaction (due to the 1:1 stoichiometry of the reaction and the very similar molar masses of the substances). We will consider N_2H_4 as the limiting reactant in the following calculations.

$$0.624 \text{ mol } N_2H_4 \times \frac{1 \text{ mol } N_2}{1 \text{ mol } N_2H_4} = 0.624 \text{ mol } N_2$$

$$0.624 \text{ mol } N_2 \times \frac{28.02 \text{ g } N_2}{1 \text{ mol } N_2} = 17.5 \text{ g } N_2$$

$$0.624 \text{ mol } N_2H_4 \times \frac{2 \text{ mol } H_2O}{1 \text{ mol } N_2H_4} = 1.248 \text{ mol } H_2O = 1.25 \text{ mol } H_2O$$

$$1.248 \text{ mol } H_2O \times \frac{18.02 \text{ g } H_2O}{1 \text{ mol } H_2O} = 22.5 \text{ g } H_2O$$

58. $12.5 \text{ g theory} \times \dfrac{40 \text{ g actual}}{100 \text{ g theory}} = 5.0 \text{ g}$

59. We are concerned with the balanced equation:

$$X_4 + 12HCl \rightarrow 4XCl_3 + 6H_2$$

We are given the mass of X_4 (248 g). If we can determine the number of moles of X_4, we can determine its molar mass, and therefore its identity. We can do this from the mass of H_2.

$$24.0 \text{ g } H_2 \times \frac{1 \text{ mol } H_2}{2.016 \text{ g } H_2} \times \frac{1 \text{ mol } X_4}{6 \text{ mol } H_2} = 1.98 \text{ mol } X_4$$

Thus,

$$\frac{248 \text{ g } X_4}{1.98 \text{ mol } X_4} = 125 \text{ g/mol (molar mass of } X_4\text{)}$$

Element X has an atomic mass of about $\frac{125}{4}$ = 31.3 g/mol. This is closest to phosphorus (30.97 g/mol).

60. We know the following:

$x + y \rightarrow xy$ and $x + 3z \rightarrow xz_3$

5 g 15 g   3 g 18 g

Thus, the mole ratio between x and y is 1:1 and the mole ratio between x and z is 1:3, respectively.

So, the relative masses of x:y are 1:3, and the relative masses of x:z are 3:6 or 1:2 (if x = 3 g, $3z$ = 18 g or z = 6 g).

Thus x:y:z = 1:3:2

If y = 60 g/mol, x = 20 g/mol and z = 40 g/mol.

61. This is a "double" limiting reactant problem. First, determine the maximum yield of P_4O_{10} from the equation:

$$P_4 + 5O_2 \rightarrow P_4O_{10}$$

$$20.0 \text{ g } P_4 \times \frac{1 \text{ mol } P_4}{123.88 \text{ g } P_4} \times \frac{1 \text{ mol } P_4O_{10}}{1 \text{ mol } P_4} \times \frac{283.88 \text{ g } P_4O_{10}}{1 \text{ mol } P_4O_{10}} = 45.8 \text{ g } P_4O_{10}$$

$$30.0 \text{ g } O_2 \times \frac{1 \text{ mol } O_2}{32.00 \text{ g } O_2} \times \frac{1 \text{ mol } P_4O_{10}}{5 \text{ mol } O_2} \times \frac{283.88 \text{ g } P_4O_{10}}{1 \text{ mol } P_4O_{10}} = 53.2 \text{ g } P_4O_{10}$$

Thus, the maximum yield of P_4O_{10} is 45.8 g.

Determine the yield of H_3PO_4 from the equation

$$P_4O_{10} + 6H_2O \rightarrow 4H_3PO_4$$

$$45.8 \text{ g } P_4O_{10} \times \frac{1 \text{ mol } P_4O_{10}}{283.88 \text{ g } P_4O_{10}} \times \frac{4 \text{ mol } H_3PO_4}{1 \text{ mol } P_4O_{10}} \times \frac{97.994 \text{ g } H_3PO_4}{1 \text{ mol } H_3PO_4} = 63.2 \text{ g } H_3PO_4$$

$$15.0 \text{ g } H_2O \times \frac{1 \text{ mol } H_2O}{18.016 \text{ g } H_2O} \times \frac{4 \text{ mol } H_3PO_4}{6 \text{ mol } H_2O} \times \frac{97.994 \text{ g } H_3PO_4}{1 \text{ mol } H_3PO_4} = 54.4 \text{ g } H_3PO_4$$

Thus, the maximum yield of H_3PO_4 is 54.4 g.

Chapter 10   Energy

1. Kinetic energy is energy of motion. Potential energy is energy of position.

2. Some energy is given off in other forms such as heat or light.

3. A state function is a function that is not dependent on the pathway. Energy is an example of a state function; heat and work are not state functions.

4. Ball A has higher potential energy in Figure 10.1a because of its higher position initially. In figure 10.1b, ball B has the higher potential energy.

5. Because energy is conserved in any process, the pathway of the process does not matter. This means that energy is a state function.

6. Heat is a flow of energy due to a temperature difference; temperature is a measure of the average kinetic energy of a substance; thermal energy comes from the random motion of the components of the substance.

7. a. No. Temperature is a measure of average kinetic energy.
   b. Yes. The thermal energy is from the random motions of the components of an object.

8. Temperature is a measure of average kinetic energy of the samples. The "coffee particles" (mostly water molecules) are of higher average kinetic energy than the "air particles" (a mixture mostly of nitrogen and oxygen molecules) in the room, which are of higher energy than the "soft drink particles" (mostly water). At the coffee-air interface a collision of a higher energy "coffee particle" with an "air particle" results in energy being transferred from the coffee to the room. Transfer also occurs from air to soft drink because the "air particles" are of higher energy than the "soft drink particles" (due to the temperature difference). The energy transfers result in the eventual average kinetic energies of each sample being equal, which means the temperatures are equal. Because the volume of air is so large (and the system is open), however, no noticeable temperature change will result.

9. There is more heat involved in mixing 100.0 g samples of 60°C and 10°C water because there is a larger temperature difference.

10. The final temperature should be less than 90°C and greater than 50°C. If the water samples had equal mass, the final temperature would be the average of the two (50°C). Because we have more of the warmer water, the final temperature will be higher than the average of the two temperatures. (Note: the actual temperature would be 63°C).

11. The potential energy is the energy available to do work. Potential energy in a chemical reaction is stored in the chemical bonds.

12. The chemical is stable; the bonds holding the atoms in the molecule are strong.

13. The reactants.

14.   a. endothermic

       b. exothermic

       c. exothermic

       d. endothermic

15. The process is exothermic; the process is endothermic.

16. The system does work on the surroundings; the surroundings does work on the system.

17. The chemist chooses the sign based on whether energy flows from the system (negative sign) or into the system (positive sign).

18. Engineers take the surroundings point of view. Thus the engineer chooses the sign based on whether energy flows from the system (positive sign) or into the system (negative sign). This way, "positive work" means the system is "doing something."

19.   a. $\Delta E = q + w = 51 \text{ kJ} + (-15 \text{ kJ}) = 36 \text{ kJ}$

     b. $\Delta E = 100. \text{ kJ} + (-65 \text{ kJ}) = 35 \text{ kJ}$

     c. $\Delta E = -65 + (-20) = -85 \text{ kJ}$

     d.   When the system delivers work to the surroundings, w < 0. This is the case in all these examples, a, b and c.

20. $\Delta E = q + w = 45 \text{ kJ} + (-29 \text{ kJ}) = 16 \text{ kJ}$

21. $\Delta E = q + w = -125 + 104 = -21 \text{ kJ}$

22. In ice, the water molecules are held together rigidly in fixed positions. As the sample is heated, the molecules vibrate but stay in their positions until the melting point is reached. Once the melting point is reached, the molecules move around much more freely (although they are still held together). As more heat is applied, the molecules move more quickly and freely until the boiling point is reached. Then the individual molecules are moving quickly enough to overcome the forces holding them together and escape from the liquid. As the molecules of the vapor are heated, they move more and more quickly.

23. lower

24. $526 \text{ J} \times \dfrac{55°C}{17°C} = 1.70 \times 10^3 \text{ J}$

25.   a. $7845 \text{ cal} \times \dfrac{4.184 \text{ J}}{1 \text{ cal}} = 3.282 \times 10^4 \text{ J} = 32.82 \text{ kJ}$

     b. $4.55 \times 10^4 \text{ cal} \times \dfrac{4.184 \text{ J}}{1 \text{ cal}} = 1.90 \times 10^5 \text{ J} = 190. \text{ kJ}$

c. $62.142 \text{ kcal} \times \dfrac{4.184 \text{ J}}{1 \text{ cal}} = 2.600 \times 10^2 \text{ kJ} = 2.600 \times 10^5 \text{ J}$

d. $43{,}024 \times \dfrac{4.184 \text{ J}}{1 \text{ cal}} = 1.800 \times 10^5 \text{ J} = 180.0 \text{ kJ}$

26. Heat = mass × specific heat capacity × temperature change
specific heat capacity = Heat/(mass × temperature change)

    72.4 kJ = 72,400 J

    specific heat capacity = $\dfrac{72{,}400 \text{ J}}{(952 \text{ g})(10.7^\circ\text{C})} = 7.11 \text{ J/g }^\circ\text{C}$

27. It is not really necessary to calculate the temperature changes experienced by the metals. For metal samples of equal mass, the metal with the smallest specific heat capacity will experience the largest temperature change when a given amount of heat is applied. Remember that the specific heat capacity represents the ability of the substance to absorb heat energy.

28. 1251 J = 35.2 g × (specific heat capacity) × 25.0°C
specific heat capacity = 1.42 J/g °C

29. Since the sign of ΔH is negative, the reaction is exothermic. Heat is evolved by the system to the surroundings.

30. $S(s) + O_2(g) \rightarrow SO_2(g) \quad \Delta H = \dfrac{-296 \text{ kJ}}{\text{mol}}$; Molar mass of SO_2 = 64.07 g/mol

    a. $275 \text{ g S} \times \dfrac{1 \text{ mol S}}{32.07 \text{ g S}} \times \dfrac{-296 \text{ kJ}}{\text{mol S}} = -2.54 \times 10^3 \text{ kJ heat released}$

    b. $25 \text{ mol S} \times \dfrac{-296 \text{ kJ}}{\text{mol S}} = -7.4 \times 10^3 \text{ kJ}$

    c. $150. \text{ g } SO_2 \times \dfrac{1 \text{ mol } SO_2}{64.07 \text{ g } SO_2} \times \dfrac{-296 \text{ kJ}}{\text{mol } SO_2} = -693 \text{ kJ}$

31. $1.00 \text{ g } CH_4 \times \dfrac{1 \text{ mol } CH_4}{16.04 \text{ g } CH_4} \times \dfrac{-891 \text{ kJ}}{\text{mol } CH_4} = -55.5 \text{ kJ}$

32.
$S + 3/2\, O_2 \rightarrow SO_3$	$\Delta H = -395.2 \text{ kJ}$
$SO_3 \rightarrow SO_2 + 1/2\, O_2$	$\Delta H = -1/2(-198.2 \text{ kJ}) = 99.1 \text{ kJ}$
$S(s) + O_2(g) \rightarrow SO_2(g)$	$\Delta H = -296.1 \text{ kJ}$

33.
$2C + 2O_2 \rightarrow 2CO_2$	$\Delta H = 2(-394 \text{ kJ})$
$H_2 + 1/2\, O_2 \rightarrow H_2O$	$\Delta H = -286 \text{ kJ}$
$2CO_2 + H_2O \rightarrow C_2H_2 + 5/2\, O_2$	$\Delta H = -(-1300.\text{ kJ})$
$2C(s) + H_2(g) \rightarrow C_2H_2(g)$	$\Delta H = 226 \text{ kJ}$

34.
$NO + O_3 \rightarrow NO_2 + O_2$	$\Delta H = -199$ kJ
$3/2\ O_2 \rightarrow O_3$	$\Delta H = -1/2(-427$ kJ$)$
$O \rightarrow 1/2\ O_2$	$\Delta H = -1/2(495$ kJ$)$
$NO(g) + O(g) \rightarrow NO_2(g)$	$\Delta H = -233$ kJ

35. To avoid fractions, let's first calculate ΔH for the reaction:

$6FeO(s) + 6CO(g) \rightarrow 6Fe(s) + 6CO_2(g)$

$6FeO + 2CO_2 \rightarrow 2Fe_3O_4 + 2CO$	$\Delta H^\circ = -2(18$ kJ$)$
$2Fe_3O_4 + CO_2 \rightarrow 3Fe_2O_3 + CO$	$\Delta H^\circ = -(-39$ kJ$)$
$3Fe_2O_3 + 9CO \rightarrow 6Fe + 9CO_2$	$\Delta H^\circ = 3(-23$ kJ$)$
$6FeO(s) + 6CO(g) \rightarrow 6Fe(s) + 6CO_2(g)$	$\Delta H^\circ = -66$ kJ

So for: $FeO(s) + CO(g) \rightarrow Fe(s) + CO_2(g)$ $\Delta H^\circ = \dfrac{-66\text{ kJ}}{6} = -11$ kJ

36. The quality of energy tells us the form of the energy (potential or kinetic). The quantity of energy tells us how much. The total amount (or quantity) is conserved. However, when potential energy is converted to kinetic energy, we say that the quality is decreasing.

37. If there is no temperature difference, there can be no flow of energy.

38. Tetraethyl lead reduced engine "knocking." It also lowered the effectiveness of catalytic converters (which lower air pollution), and increased the amount of lead in the environment.

39. If it were not for greenhouse gases the surface temperature of the earth would be much colder. Having too large a quantity of greenhouse gases may cause the surface temperature of the earth to rise to dangerous levels.

40. Wood has decreased the most. Nuclear and petroleum/natural gas have increased the most (see Figure 10.7).

41. The first law of thermodynamics tells us that the total amount of energy is constant. It does not tell us anything about direction of energy transfer.

42. A driving force is a force that tends to make a process occur. We know, for example, that when we open a pressurized tank of gas, the gas will move from the tank to the surroundings. Moving from high to low pressure is a driving force for a gas.

43. The driving force of an exothermic reaction is energy spread. In an exothermic reaction energy is transferred from the system to the surroundings.

44. The energy of the surroundings is used to melt solid water. The temperature of the water will eventually reach that of the room, and room temperature is higher than the melting point of water. In addition, liquid water has greater entropy than solid water, so entropy is the driving force in this process.

Energy

45. As the steam is cooled from 150°C to 100°C, the molecules of vapor gradually slow down as they lose kinetic energy. At 100°C, the steam condenses into liquid water, and the temperature remains at 100°C until all the steam has condensed. As the liquid water cools, the molecules in the liquid move more and more slowly as they lose kinetic energy. At 0°C, the liquid water freezes.

46. Since 1.000 cal = 4.184 J, then 1.000 kcal = 4.184 kJ

   a. $462.4 \text{ kJ} \times \dfrac{1 \text{ kcal}}{4.184 \text{ kJ}} = 110.5 \text{ kcal}$

   b. $18.28 \text{ kJ} \times \dfrac{1 \text{ kcal}}{4.184 \text{ kJ}} = 4.369 \text{ kcal}$

   c. $1.014 \text{ kJ} \times \dfrac{1 \text{ kcal}}{4.184 \text{ kJ}} = 0.2424 \text{ kcal}$

   d. $190.5 \text{ kJ} \times \dfrac{1 \text{ kcal}}{4.184 \text{ kJ}} = 45.53 \text{ kcal}$

47. a. $45.62 \text{ kcal} \times \dfrac{4.184 \text{ kJ}}{1 \text{ kcal}} = 190.9 \text{ kJ}$

   b. $72.94 \text{ kJ} \times \dfrac{1 \text{ kcal}}{4.184 \text{ kJ}} = 17.43 \text{ kcal}$

   c. $2.751 \text{ kJ} \times \dfrac{1 \text{ kcal}}{4.184 \text{ kJ}} \times \dfrac{1000 \text{ cal}}{1 \text{ kcal}} = 657.5 \text{ cal}$

   d. $5.721 \text{ kcal} \times \dfrac{4.184 \text{ kJ}}{1 \text{ kcal}} \times \dfrac{1000 \text{ J}}{1 \text{ kJ}} = 2.394 \times 10^4 \text{ J}$

48. Temperature increase = 75.0 − 22.3 = 52.7°C

   $145 \text{ g} \times 4.184 \text{ J/g °C} \times 52.7°\text{C} \times \dfrac{1 \text{ cal}}{4.184 \text{ J}} = 7641.5 \text{ cal} = 7.64 \times 10^3 \text{ cal}$

49. Table 10.1 gives the specific heat of silver as 0.24 J/g °C

   Temperature increase = 15.2 − 12.0 = 3.2°C

   1.25 kJ = 1250 J

   1250 J = (mass of silver) × 0.24 J/g °C × 3.2°C

   mass of silver = 1627 g = 1.6×10^3 g silver

50. Table 10.1 gives the specific heat capacity of iron as 0.45 J/g°C

   50. joules is the heat that is applied to the sample of iron, and must equal the product of the mass of iron, the specific heat capacity of the iron, and the temperature change undergone by the iron (which is what we want).

Heat = mass x specific heat capacity x temperature change

50. J = 10. g x 0.45 J/g °C x ΔT

$$\Delta T = \frac{50. \text{ J}}{10. \text{ g} \times 0.45 \text{ J/g °C}} = 11°C$$

51. $0.13 \dfrac{\text{J}}{\text{g °C}} \times \dfrac{1 \text{ cal}}{4.184 \text{ J}} = 0.031 \dfrac{\text{cal}}{\text{g °C}}$

52. 2.5 kg of water = 2,500 g

    Temperature change = 55.0 – 18.5 = 36.5°C

    2500 g x 4.184 J/g °C x 36.5°C = 3.8 x 10^5 J

53. For a given mass of substance, the substance with the *smallest* specific heat capacity (gold, 0.13 J/ °C) will undergo the *largest* increase in temperature. Conversely, the substance with the largest specific heat capacity (water, 4.184 J/g °C) will undergo the smallest increase in temperature.

54. Let T_f represent the final temperature reached.

    For the hot water, heat lost = 50.0 g x 4.184 J/g °C x (100. – T_f)

    For the cold water, heat gained = 50.0 g x 4.184 J/g °C x (T_f – 25)
    Heat lost by the hot water must *equal* heat gained by the cold water.

    50.0 g x 4.184 J/g °C x (100. – T_f) = 50.0 g x 4.184 J/g °C x (T_f – 25)

    209.2(100 – T_f) = 209.2(T_f – 25)

    (100 – T_f) = (T_f – 25)

    2 T_f = 125°C

    T_f = 62.5°C = 63°C

55. $(m \times s \times \Delta T)_{water} = (m \times s \times \Delta T)_{iron}$

    Let T_f represent the final temperature reached by the system

    [75 g x 4.184 J/g °C x (T_f – 20.)] = [25.0 g x 0.45 J/g °C x (85 – T_f)]

    314(T_f – 20.) = 11.4(85 – T_f)

    314T_f – 6280 = 961 – 11.3T_f

    325T_f = 7241

    T_f = 22.3°C = 22°C

56. 9.0 J (It requires twice as much heat to warm twice as large a sample over the same temperature interval.)

57. For any substance, $Q = m \times s \times \Delta T$. The basic calculation for each of the substances is the same (specific heat capacities are found in Table 10.1).

Heat required = 150. g × (specific heat capacity) × 11.2°C

Substance	Specific Heat Capacity	Heat Required
water (l)	4.184 J/g °C	7.03 × 10³ J
water (s)	2.03 J/g °C	3.41 × 10³ J
water (g)	2.0 J/g °C	3.4 × 10³ J
aluminum	0.89 J/g °C	1.5 × 10³ J
iron	0.45 J/g °C	7.6 × 10² J
mercury	0.14 J/g °C	2.4 × 10² J
carbon	0.71 J/g °C	1.2 × 10³ J
silver	0.24 J/g °C	4.0 × 10² J
gold	0.13 J/g °C	2.2 × 10² J

58. Since, for any substance, $Q = m \times s \times \Delta T$, we can solve this equation for the temperature change, ΔT. The results are tabulated:

Substance	Specific Heat Capacity	Temperature Change
water (l)	4.184 J/g °C	23.9°C
water (s)	2.03 J/g °C	49.3°C
water (g)	2.0 J/g °C	50.°C
aluminum	0.89 J/g °C	1.1 × 10² °C
iron	0.45 J/g °C	2.2 × 10² °C
mercury	0.14 J/g °C	7.1 × 10² °C
carbon	0.71 J/g °C	1.4 × 10² °C
silver	0.24 J/g °C	4.2 × 10² °C
gold	0.13 J/g °C	7.7 × 10² °C

59.   a. $\Delta E = q + w = -47$ kJ $+ 88$ kJ $= 41$ kJ

    b. $\Delta E = 82 + 47 = 129$ kJ

    c. $\Delta E = 47 + 0 = 47$ kJ

    d.   When the surroundings deliver work to the system, w > 0. This is the case for a and b.

60.   a.   The combustion of gasoline releases heat, so this is an exothermic process.

    b. $H_2O(g) \rightarrow H_2O(l)$; Heat is released when water vapor condenses, so this is an exothermic process.

    c.   To convert a solid to a gas, heat must be absorbed, so this is an endothermic process.

    d.   Heat must be added (absorbed) in order to break a bond, so this is an endothermic process.

61. $4Fe(s) + 3O_2(g) \rightarrow 2Fe_2O_3(s)$ $\Delta H = -1652$ kJ; Note that 1652 kJ of heat are released when 4 mol Fe reacts with 3 mol O_2 to produce 2 mol Fe_2O_3.

    a.   4.00 mol Fe × $\dfrac{-1652 \text{ kJ}}{4 \text{ mol Fe}} = -1650$ kJ

b. $1.00 \text{ ml Fe}_2\text{O}_3 \times \dfrac{-1652 \text{ kJ}}{2 \text{ mol Fe}_2\text{O}_3} = -826 \text{ kJ}$

c. $1.00 \text{ g Fe} \times \dfrac{1 \text{ mol Fe}}{55.85 \text{ g}} \times \dfrac{-1652 \text{ kJ}}{4 \text{ mol Fe}} = -7.39 \text{ kJ}$

d. $10.0 \text{ g Fe} \times \dfrac{1 \text{ mol Fe}}{55.85 \text{ g}} = 0.179 \text{ mol Fe}$; $2.00 \text{ g O}_2 \times \dfrac{1 \text{ mol O}_2}{32.00 \text{ g}} = 0.0625 \text{ mol O}_2$

0.179 mol Fe/0.0625 mol O_2 = 2.86; The balanced equation requires a 4 mol Fe/3 mol O_2 = 1.33 mol ratio. O_2 is limiting since the actual mol Fe/mol O_2 ratio is less than the required mol ratio.

$0.0625 \text{ mol O}_2 \times \dfrac{-1652 \text{ kJ}}{3 \text{ mol O}_2} = -34.4 \text{ kJ}$ heat released

62. Reversing the first equation and multiplying by 1/6, we get

$3/6 \text{ D} \rightarrow 3/6 \text{ A} + \text{B}$ $\Delta H = \dfrac{+403 \text{ kJ / mol}}{6}$

or

$1/2 \text{ D} \rightarrow 1/2 \text{ A} + \text{B}$ $\Delta H = +67.2 \text{ kJ/mol}$

Dividing the second equation by 2, we get

$1/2 \text{ E} + \text{F} \rightarrow 1/2 \text{ A}$ $\Delta H = \dfrac{-105.2 \text{ kJ / mol}}{2} = -52.6 \text{ kJ/mol}$

Dividing the third equation by 2, we get

$1/2 \text{ C} \rightarrow 1/2 \text{ E} + 3/2 \text{ D}$ $\Delta H = \dfrac{+64.8 \text{ kJ / mol}}{2} = 32.4 \text{ kJ/mol}$

Adding these, we get

$1/2 \text{ D} \rightarrow 1/2 \text{ A} + \text{B}$

$1/2 \text{ E} + \text{F} \rightarrow 1/2 \text{ A}$

$1/2 \text{ C} \rightarrow 1/2 \text{ E} + 3/2 \text{ D}$

$1/2 \text{ E} + 1/2 \text{ D} + 1/2 \text{ C} + \text{F} \rightarrow 1/2 \text{ A} + 1/2 \text{ A} + 1/2 \text{ E} + 3/2 \text{ D} + \text{B}$

or

$1/2 \text{ C} + \text{F} \rightarrow \text{A} + \text{B} + \text{D}$ $\Delta H = 47.0 \text{ kJ/mol}$

63. During exercise, the body generates about 5500 kJ/hr, or about 11,000 kJ in 2 hours. Assuming all of this heat is lost through evaporation, we can calculate the volume of the perspiration. Water has a heat of vaporization of 40.7 kJ/mol. So,

$11{,}000 \text{ kJ} \times \dfrac{1 \text{ mol H}_2\text{O}}{40.7 \text{ kJ}} = 270 \text{ mol H}_2\text{O}$

Energy

$$270 \text{ mol H}_2\text{O} \times \frac{18.016 \text{ g H}_2\text{O}}{1 \text{ mol H}_2\text{O}} = 4900 \text{ g H}_2\text{O}$$

Assuming a density of 1 g/mL for H_2O, we get

$$4900 \text{ g H}_2\text{O} \times \frac{1 \text{ mL H}_2\text{O}}{1 \text{ g H}_2\text{O}} \times \frac{1 \text{ L}}{1000 \text{ mL}} = 4.9 \text{ L H}_2\text{O}$$

Thus, almost 5 liters (almost 1 1/3 gallons) of water must be evaporated by perspiration.

64. If 400 kcal is burned per hour walking 4.0 mph, then the total amount of heat burned while walking at 4.0 mph will be given by

$$Q = 400 \text{ kcal/hr} \times t$$

Where t is the amount of time spent walking.
One gram of fat is consumed for every 7.7 kcal of heat. Therefore, one pound of fat requires

$$7.7 \text{ kcal/g} \times 454 \text{ g/lb} = 3500 \text{ kcal/lb}$$

Since we want to lose one pound, $Q = 3500$ kcal, and

$$t = \frac{Q}{400 \text{ kcal/hr}} = \frac{3500 \text{ kcal}}{400 \text{ kcal/hr}} = 8.75 \text{ hr} \; (\approx 9 \text{ hr})$$

Chapter 11   Modern Atomic Theory

1. This experiment is explained in Chapter 3.

2. See Figure 11.1 for a sketch of Rutherford's atom. Rutherford could not answer questions such as "Why aren't the negative electrons attracted into the positive nucleus, causing the atom to collapse?"

3. Electromagnetic radiation is radiant energy that travels through space with wavelike behavior. There are many examples of electromagnetic radiation: light, radio/television signals, microwaves, heat, X-rays, cosmic rays, etc.

4. The different forms of electromagnetic radiation are similar in that they all exhibit the same type of wave-like behavior and are propagated through space at the same speed (the speed of light). The types of electromagnetic radiation differ in their frequency (and wavelength) and in the resulting amount of energy carried per photon.

5. The *wavelength* represents the distance between two corresponding points (peaks, troughs, etc.) on successive cycles of a wave. Figure 11.3 in the text illustrates a typical wave. The *frequency* of electromagnetic radiation represents how many complete cycles of the wave pass a given point per second. The wavelength and frequency are related, however, since obviously a longer wave will take more time to pass a given point in space than a shorter wave, if the two waves are propagated at the same speed.

6. The *speed* of electromagnetic radiation represents how fast a given wave moves through space. The *frequency* of electromagnetic radiation represents how many complete cycles of the wave pass a given point per second. For example, your favorite radio station broadcasts waves of a particular frequency that distinguishes it, but how long those waves take to reach you depends on their speed through space.

7. photon

8. Although electromagnetic radiation exhibits characteristic wave-like properties, such radiation also demonstrates a particle-like nature. For example, when an excited atom emits radiation, the radiation demonstrates wave properties, but occurs in discrete particle-like bundles (photons). The wave-particle nature of light refers to the fact that a beam of electromagnetic energy can be considered not only as a continuous wave, but also as a stream of discrete packets of energy moving through space.

9. The colors are due to the releasing of energy.

10. Higher energy photons are released with Cu^{2+} because green light is emitted (red light is emitted from Li^+). Green light is of higher energy than red light.

11. An atom is said to be in an excited state when it possesses more than its minimum energy (its ground state). An atom is promoted from its ground state to an excited state by the absorption of energy, and returns to its ground state by the emission of the excess energy.

12. A photon having an energy corresponding to the energy difference between the two states is emitted by an atom in an excited state when it returns to its ground state.

Modern Atomic Theory

13. The emission of light by excited atoms has been the key interconnection between the macroscopic world we can observe and measure, and with what is happening on a microscopic basis within an atom. Excited atoms emit light (which we can measure) because of changes in the microscopic structure of the atom. By studying the emissions of atoms we can trace back to what happened inside the atom.

14. When excited hydrogen atoms emit their excess energy, the photons of radiation emitted are always of exactly the same wavelength and energy. We consider this to mean that the hydrogen atom possesses only certain allowed energy states, and that the photons emitted correspond to the atom changing from one of these allowed energy states to another of the allowed energy state. The energy of the photon emitted corresponds to the energy difference in the allowed states. If the hydrogen atom did not possess discrete energy levels, then we would expect the photons emitted to have random wavelengths and energies.

15. Hydrogen always emits light at exactly the same wavelengths, corresponding to transitions of the electron between the fixed energy states within the atom.

16. The ground state of an atom is its lowest possible energy state.

17. Bohr pictured electrons moving in circular orbits corresponding to the various allowed energy levels. He suggested that the electron could jump to a different orbit by absorbing or emitting a photon of light with exactly the correct energy content (corresponding to the difference in energy between the orbits).

18. According to Bohr, electrons move in discrete, fixed circular *orbits* around the nucleus. If the wavelength of the applied energy corresponds to the *difference in energy* between the two orbits, the atom absorbs a photon and the electron moves to a larger orbit.

19. Bohr suggested that the electron could jump to a different orbit by absorbing or emitting a photon of light with exactly the correct energy content (corresponding to the difference in energy between the orbits). Since the energy levels of a given atom were fixed and definite, then the atom should always emit energy at the same discrete wavelengths.

20. Bohr's theory *explained* the experimentally *observed* line spectrum of hydrogen *exactly*. Bohr's theory was ultimately discarded because when attempts were made to extend the theory to atoms other than hydrogen, the calculated properties did *not* correspond closely to experimental measurements.

21. Schrödinger and de Broglie reasoned that, since light seems to have both wave and particle characteristics (it behaves simultaneously as a wave and as if it were a stream of particles), that perhaps the electron might exhibit both of these characteristics. That is, although the electron behaves as a discrete particle, perhaps the properties of the electron in the atom could be treated as if they were wavelike.

22. An orbit represents a definite, exact circular pathway around the nucleus in which an electron can be found. An orbital represents a region of space in which there is a high probability of finding the electron.

23. Any experiment which sought to measure the exact location of an electron (such as shooting a beam of light at it) would cause the electron to move. Any measurement made

would necessitate the application or removal of energy, which would disturb the electron from where it had been before the measurement.

24. Chemists arbitrarily *defined* an orbital to represent a 90% probability of finding the electron within this region. Orbitals represent mathematical functions and have no distinct outer edge (so they are drawn to appear "fuzzy": they are *not* hard-edged capsules enclosing the electron, but merely represent the most likely region where an electron may be found.)

25. Pictures we draw to represent orbitals should only be interpreted as probability maps. They are not meant to represent that the electron moves only on the surface of, or within, the region drawn in the picture. Since the mathematical probability of finding the electron never actually becomes zero on moving outward from the nucleus, scientists have decided that pictures of orbitals should represent a 90% probability that the electron will be found inside the region depicted in the drawing (for 100% probability, the orbital would have to encompass all space).

26. The 2s orbital is similar in shape to the 1s orbital, but is larger.

27. The higher the principal energy level (n), the farther from the nucleus, on average, the electron will be.

28. increases; the energy of an electron increases with its principal quantum number.

29. The other orbitals serve as the excited states of the hydrogen atom. When energy of the right frequency is applied to the hydrogen atom, the electron can move from its normal orbital (ground state) to one of the other orbitals (excited states). Later on, the electron can move back to its normal orbital and release the absorbed energy as light.

30. The Pauli exclusion principle states that an orbital can hold a maximum of two electrons, and those two electrons must have opposite spins.

31. The higher the value of the principal quantum number, n, the higher the energy of the principal energy level.

32. increases; as you move out from the nucleus, there is more space and room for more sublevels.

33. a. correct (the $n = 2$ shell contains s and p subshells)

    b. incorrect (the $n = 1$ shell consists only of $1s$ orbitals)

    c. incorrect (the $n = 3$ shell contains $3s$, $3p$, and $3d$ orbitals)

    d. correct (the $n = 4$ shell contains s, p, d, and f subshells)

34. The 1s orbital is closest to the nucleus and lowest in energy, so it is always filled first.

35. Valence electrons are those in the outermost (highest) principal energy level of an atom. These electrons are especially important because they are at the "outside edge" of an atom, and are those electrons which are "seen" by other atoms and which can interact with the electrons of another atom in a chemical reaction.

Modern Atomic Theory

36. The elements in a given vertical column of the periodic table have the same valence electron configuration. Having the same valence electron configuration causes the elements in a given group to have similar chemical properties.

37.
a. $1s^2\ 2s^2\ 2p^6\ 3s^2\ 3p^6\ 4s^2\ 3d^{10}\ 4p^6\ 5s^2$
b. $1s^2\ 2s^2\ 2p^6\ 3s^2\ 3p^6\ 4s^2\ 3d^{10}$
c. $1s^2$
d. $1s^2\ 2s^2\ 2p^6\ 3s^2\ 3p^6\ 4s^2\ 3d^{10}\ 4p^5$

38.
a. $1s^2\ 2s^2\ 2p^6\ 3s^2\ 3p^6\ 4s^2$
b. $1s^2\ 2s^2\ 2p^6\ 3s^2\ 3p^6\ 4s^1$
c. $1s^2\ 2s^2\ 2p^5$
d. $1s^2\ 2s^2\ 2p^6\ 3s^2\ 3p^6\ 4s^2\ 3d^{10}\ 4p^6$

39.
a. (↑↓) (↑↓) (↑↓)(↑↓)(↑↓) (↑↓) (↑)( )( )
   1s   2s   2p   3s   3p

b. (↑↓) (↑↓) (↑↓)(↑↓)(↑↓) (↑↓) (↑)(↑)(↑)
   1s   2s   2p   3s   3p

c. (↑↓) (↑↓) (↑↓)(↑↓)(↑↓) (↑↓) (↑↓)(↑↓)(↑↓) (↑↓)
   1s   2s   2p   3s   3p   4s
   (↑↓)(↑↓)(↑↓)(↑↓)(↑↓) (↑↓)(↑↓)(↑)
          3d   4p

d. (↑↓) (↑↓) (↑↓)(↑↓)(↑↓) (↑↓) (↑↓)(↑↓)(↑↓)
   1s   2s   2p   3s   3p

40. For the representative elements (those filling s and p subshells), the group number gives the number of valence electrons.

a. one ($3s$)
b. two ($4s$)
c. seven ($5s, 5p$)
d. five ($2s, 2p$)

41. This belief is based on the *experimental properties* of K and Ca. The physical and chemical properties of K are like those of the other Group 1 elements; Ca's properties are similar to the other Group 2 elements.

42. The properties of Rb and Sr suggest that they are members of Groups 1 and 2, respectively, and so must be filling the 5s orbital. The 5s orbital is lower in energy (and fills before) the 4d orbitals.

43.
a. [Ar] $4s^2$
b. [Rn] $7s^1$

c. [Kr] $5s^2\, 4d^1$

d. [Xe] $6s^2\, 4f^1\, 5d^1$

44. a. [Ne] $3s^2\, 3p^3$

b. [Ne] $3s^2\, 3p^5$

c. [Ne] $3s^2$

d. [Ar] $4s^2\, 3d^{10}$

45. a. one

b. two

c. zero

d. ten

46. Figure 10.27 shows the orbitals being filled as a function of location in the periodic table.

a. $5f$

b. $5f$

c. $4f$

d. $6p$

47. Figure 10.30 shows partial electronic configurations.

a. [Rn] $7s^2\, 5f^3\, 6d^1$

b. [Ar] $4s^2\, 3d^5$

c. [Xe] $6s^2\, 4f^{14}\, 5d^{10}$

d. [Rn] $7s^1$

48. The metallic elements *lose* electrons and form *positive* ions (cations); the nonmetallic elements *gain* electrons and form *negative* ions (anions). Remember that the electron itself is *negatively* charged.

49. The Group 1 metals are all highly reactive, and all form 1+ ions almost exclusively when they react. Physically, these metals are soft (they can be cut with a knife) and very low in density. Because of their high reactivity, these metals tend to be found with a coating of the metal oxide which hides their metallic luster (which can be seen, however, if a fresh surface of the metal is exposed).

50. All exist as *diatomic* molecules (F_2, Cl_2, Br_2, I_2); all are *non*metals; all have relatively high electronegativities; all form 1– ions in reacting with metallic elements.

51. The nonmetallic elements are clustered at the upper right side of the periodic table. These elements are effective at pulling electrons from metallic elements for several reasons. First, these elements have little tendency to lose electrons themselves (they have high ionization energies). Secondly, the atoms of these elements tend to be small in size, which means that electrons can be pulled in strongly since they can get closer to the

Modern Atomic Theory

nucleus. Finally, if these atoms gain electrons, they can approach the electronic configuration of the following noble gas elements (see Chapter 12 for why the electronic configuration of the noble gases are desirable for other atoms to attain).

52. The elements of a given period (horizontal row) have valence electrons in the same principal energy level. Nuclear charge, however, increases across a period going from left to right. Atoms at the left side have smaller nuclear charges, and hold onto their valence electrons less tightly.

53. All atoms within a given group have the same number of valence electrons, and these valence electrons occur in the same type of subshell. However, at the bottom of a group, the valence electrons are in a higher principal energy level. The *principal* energy levels increase in distance from the nucleus as the principal quantum number, *n*, increases.

54. The *nuclear charge* increases from left to right within a period, pulling progressively more tightly on the valence electrons.

55. For most elements, the chemical activity is reflected in the ease with which the element gains or loses electrons.
    a. Na (the less reactive metals are further up in a group)
    b. Be (the less reactive metals are further up in a group)
    c. Br (the less reactive nonmetals are at the bottom of a group)
    d. Te (the less reactive nonmetals are at the bottom of a group)

56. Ionization energies decrease in going from top to bottom within a vertical group; ionization energies increase in going from left to right within a horizontal period.
    a. Li
    b. Ca
    c. Cl
    d. S

57. Atomic size increases in going from top to bottom within a vertical group; atomic size decreases in going from left to right within a horizontal period.
    a. Xe < Sn < Sr < Rb
    b. He < Kr < Xe < Rn
    c. At < Pb < Ba < Cs

58. wavelength

59. speed of light

60. quantized

61. orbital

62. a. $1s^2\,2s^2\,2p^6\,3s^2\,3p^6\,4s^1$ [Ar] $4s^1$

(↑↓) (↑↓) (↑↓)(↑↓)(↑↓) (↑↓) (↑↓)(↑↓)(↑↓) (↑)
 1s   2s   2p   3s   3p   4s

b. $1s^2\,2s^2\,2p^6\,3s^2\,3p^6\,4s^2\,3d^2$ [Ar] $4s^2\,3d^2$

(↑↓) (↑↓) (↑↓)(↑↓)(↑↓) (↑↓) (↑↓)(↑↓)(↑↓) (↑↓) (↑)(↑)( )( )( )
 1s   2s   2p   3s   3p   4s   3d

c. $1s^2\,2s^2\,2p^6\,3s^2\,3p^2$ [Ne] $3s^2\,3p^2$

(↑↓) (↑↓) (↑↓)(↑↓)(↑↓) (↑↓) (↑)(↑)( )
 1s   2s   2p   3s   3p

d. $1s^2\,2s^2\,2p^6\,3s^2\,3p^6\,4s^2\,3d^6$ [Ar] $4s^2\,3d^6$

(↑↓) (↑↓) (↑↓)(↑↓)(↑↓) (↑↓) (↑↓)(↑↓)(↑↓) (↑↓) (↑↓)(↑)(↑)(↑)(↑)
 1s   2s   2p   3s   3p   4s   3d

e. $1s^2\,2s^2\,2p^6\,3s^2\,3p^6\,4s^2\,3d^{10}$ [Ar] $4s^2\,3d^{10}$

(↑↓) (↑↓) (↑↓)(↑↓)(↑↓) (↑↓) (↑↓)(↑↓)(↑↓) (↑↓) (↑↓)(↑↓)(↑↓)(↑↓)(↑↓)
 1s   2s   2p   3s   3p   4s   3d

63. a. ns^2

b. $ns^2\,np^5$

c. $ns^2\,np^4$

d. ns^1

e. $ns^2\,np^4$

64. a. four (two if the *d* electrons are not counted as valence electrons)

b. seven

c. two

d. seven (two if the *d* electrons are not counted as valence electrons)

65. Light is emitted from the hydrogen atom only at certain fixed wavelengths. If the energy levels of hydrogen were *continuous*, a hydrogen atom would emit energy at all possible wavelengths.

66. The three 2*p* orbitals of carbon are of the *same energy*; by occupying different orbitals of the same energy, repulsion between electrons is minimized.

67. a. $1s^2\,2s^2\,2p^6\,3s^2\,3p^6\,4s^2\,3d^{10}\,4p^5$

b. $1s^2\,2s^2\,2p^6\,3s^2\,3p^6\,4s^2\,3d^{10}\,4p^6\,5s^2\,4d^{10}\,5p^6$

c. $1s^2\,2s^2\,2p^6\,3s^2\,3p^6\,4s^2\,3d^{10}\,4p^6\,5s^2\,4d^{10}\,5p^6\,6s^2$

d. $1s^2\,2s^2\,2p^6\,3s^2\,3p^6\,4s^2\,3d^{10}\,4p^4$

Modern Atomic Theory

68. metals, low; nonmetals, high

69. Atomic size increases in going from top to bottom within a vertical group; atomic size decreases in going from left to right within a horizontal period.

   a. Ca
   b. P
   c. K

70. a. The number of protons = (mass number) – (number of neutrons)
   $$= 203 - 123$$
   $$= 80 \text{ protons}$$

   b. The element with 80 protons is mercury (Hg)
   c. 2 valence electrons (in the 6s orbital)
   d. There are six levels of electrons.

71. The elements are

   i) Mg; ii) Na; iii) Ne; iv) O; v) N; vi) B

   a. iv (O)
   b. ii (Na)
   c. ii (Na)
   d. v (N, with 3 unpaired electrons)
   e. i (Mg)
   f. Na$_3$N (Na$^+$, N^{3-})
   g. Mg$_3$N$_2$ (Mg^{2+}, N^{3-})
   h. ii, i, vi, v, iv, iii

72. a. F
   b. Lu
   c. Po
   d. Fr
   e. At
   f. Xe
   g. Lr

73. Element 120 would be a solid and a metal. It would be expected to be highly reactive with water and have a tendency to form ions with a 2+ charge.

74. Elements in the same column (family) generally have similar properties. Thus, we predict the following families:

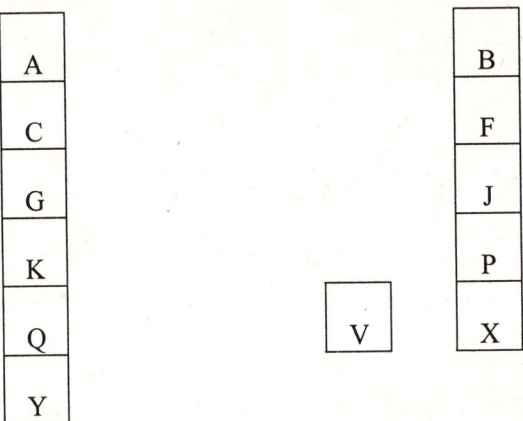

We order the remaining elements by atomic mass. There are many possible answers. The most compact table would have the 5th period filled as follows:

A							B
C							F
G							J
K							P
Q	R	S	T	U	V	W	X
Y							

The other elements could then exist in the following spots:

A							B
C							F
G							J
K							P
Q	R	S	T	U	V	W	X
Y							

Chapter 12   Chemical Bonding

1. A chemical bond represents a force which holds groups of two or more atoms together and makes them function as a unit.

2. An ionic compound results when a metallic element reacts with a nonmetallic element. An example is the reaction of the metal sodium with the nonmetal chlorine:

    $$2Na(s) + Cl_2(g) \rightarrow 2Na^+Cl^-(s)$$

3. *Ionic* bonding results from the complete transfer of an electron from one atom to another, whereas *covalent* bonding exists when two atoms share pairs of electrons.

4. The H_2 molecule contains two atoms of the same element. When two hydrogen atoms are brought close together, each hydrogen atom contributes its electron to the formation of a covalent bond, with the resulting pair of electrons shared equally by the two hydrogen atoms. An H_2 molecule is more stable than two separated hydrogen atoms. The pair of electrons is simultaneously attracted by the two nuclei.

5. The HF molecule contains atoms of two different elements. In bonding with each other, these atoms share a pair of valence electrons, but they do not share them equally. The bonding in HF is described as polar covalent, which indicates that although valence electron pairs are shared between atoms, the electron pair is attracted more strongly by one of the atoms than the other. Nonetheless, an HF molecule is more stable in energy than are separated H and F atoms.

6. Electronegativity is the ability of an atom to attract a shared pair of electrons toward itself in forming a chemical bond. If we postulate a bond forming between two atoms, the relative electronegativies of the atoms will tell us what type of chemical bond is likely to be formed.

7. A polar covalent bond results when one atom of the bond attracts electrons more strongly toward itself than does the second atom of the bond. A polar covalent bond results when the atoms forming the covalent bond have different electronegativities. The fact that the bonds in a molecule are polar does *not* necessarily mean that the overall molecule itself will be polar: the overall polarity of the molecule also depends on the geometry of the molecule.

8. In each case, the element higher up within a given group of the periodic table, or to the right within a period, has the higher electronegativity.

    a. K < Ca < Sc

    b. At < Br < F

    c. C < N < O

9. Generally, covalent bonds between atoms of *different* elements are *polar*.

    a. covalent

    b. polar covalent

c. polar covalent

d. ionic

10. For a bond to be polar covalent, the atoms involved in the bond must have different electronegativities (must be of different elements).

   a. *non*polar covalent (atoms of the same element)

   b. *non*polar covalent (atoms of the same element)

   c. *non*polar covalent (atoms of the same element)

   d. polar covalent (atoms of different elements)

11. The *degree* of polarity of a polar covalent bond is indicated by the magnitude of the difference in electronegativities of the elements involved: the larger the difference in electronegativity, the more polar is the bond. Electronegativity differences are given in parentheses below:

   a. H–O (1.4); H–N (0.9); the H–O bond is more polar.

   b. H–N (0.9); H–F (1.9); the H–F bond is more polar.

   c. H–O (1.4); H–F (1.9); the H–F bond is more polar.

   d. H–O (1.4); H–Cl (0.9); the H–O bond is more polar.

12. The greater the electronegativity difference between two atoms, the more ionic will be the bond between those two atoms. Electronegativity differences are given in parentheses.

   a. Na–O (2.6) has more ionic character than Na–N (2.1)

   b. K–S (1.7) has more ionic character than K–P (1.3)

   c. K–Cl (2.2) has more ionic character than Na–Cl (2.1)

   d. Na–Cl (2.1) has more ionic character than Mg–Cl (1.8)

13. A dipole moment is an electrical effect that occurs in a molecule that has separate centers of positive and negative charge. The simplest examples of molecules with dipole moments would be diatomic molecules involving two different elements. For example:
   δ+ C→O δ– δ+ N→O δ– δ+ Cl→F δ– δ+ Br→Cl δ–

14. The presence of strong bond dipoles and a large overall dipole moment in water make it a very polar substance overall. Among those properties of water that are dependent on its dipole moment are its freezing point, melting point, vapor pressure, and its ability to dissolve many substances.

15. In a diatomic molecule containing two different elements, the more electronegative atom will be the negative end of the molecule, and the *less* electronegative atom will be the positive end.

   a. H

   b. Cl

   c. I

Chemical Bonding

16. In the figures, the arrow points toward the more electronegative atom.
    a. δ+ P→F δ−
    b. δ+ P→O δ−
    c. δ+ P→C δ−
    d. P and H have similar electronegativities.

17. In the figures, the arrow points toward the more electronegative atom.
    a. δ+ P→S δ−
    b. δ+ S→O δ−
    c. δ+ S→N δ−
    d. δ+ S→Cl δ−

18. gaining

19. Atoms in covalent molecules gain a configuration like that of a noble gas by sharing one or more pairs of electrons between atoms: such shared pairs of electrons "belong" to each of the atoms of the bond at the same time. In ionic bonding, one atom completely gives over one or more electrons to another atom, and then the resulting ions behave independently of one another.

20.
    a. Li $1s^2 \, 2s^1$
       Li$^+$ $1s^2$
       He has the same configuration as Li$^+$

    b. Br $1s^2 \, 2s^2 \, 2p^6 \, 3s^2 \, 3p^6 \, 4s^2 \, 3d^{10} \, 4p^5$
       Br$^-$ $1s^2 \, 2s^2 \, 2p^6 \, 3s^2 \, 3p^6 \, 4s^2 \, 3d^{10} \, 4p^6$
       Kr has the same configuration as Br$^-$

    c. Cs $1s^2 \, 2s^2 \, 2p^6 \, 3s^2 \, 3p^6 \, 4s^2 \, 3d^{10} \, 4p^6 \, 5s^2 \, 4d^{10} \, 5p^6 \, 6s^1$
       Cs$^+$ $1s^2 \, 2s^2 \, 2p^6 \, 3s^2 \, 3p^6 \, 4s^2 \, 3d^{10} \, 4p^6 \, 5s^2 \, 4d^{10} \, 5p^6$
       Xe has the same configuration as Cs$^+$

    d. S $1s^2 \, 2s^2 \, 2p^6 \, 3s^2 \, 3p^4$
       S^{2-} $1s^2 \, 2s^2 \, 2p^6 \, 3s^2 \, 3p^6$
       Ar has the same configuration as S^{2-}

    e. Mg $1s^2 \, 2s^2 \, 2p^6 \, 3s^2$
       Mg^{2+} $1s^2 \, 2s^2 \, 2p^6$
       Ne has the same configuration as Mg^{2+}

21. a. Ca^{2+} (Ca has two electrons more than the noble gas Ar)
     b. N^{3-} (N has three electrons fewer than the noble gas Ne)
     c. Br^- (Br has one electron fewer than the noble gas Kr)
     d. Mg^{2+} (Mg has two electrons more than the noble gas Ne)

22. a. Na_2S — Na has one electron more than a noble gas; S has two electrons fewer than a noble gas.
     b. $BaSe$ — Ba has two electrons more than a noble gas; Se has two electrons fewer than a noble gas.
     c. $MgBr_2$ — Mg has two electrons more than a noble gas; Br has one electron less than a noble gas.
     d. Li_3N — Li has one electron more than a noble gas; N has three electrons fewer than a noble gas.
     e. KH — K has one electron more than a noble gas; H has one electron less than a noble gas.

23. a. Al^{3+}, [Ne]; S^{2-}, [Ar]
     b. Mg^{2+}, [Ne]; N^{3-}, [Ne]
     c. Rb^+, [Kr]; O^{2-}, [Ne]
     d. Cs^+, [Xe]; I^-, [Xe]

24. The formula of an ionic compound represents only the smallest whole number ratio of the positive and negative ions present (the empirical formula).

25. An ionic solid such as NaCl basically consists of an array of alternating positively and negatively charged ions: that is, each positive ion has as its nearest neighbors a group of negative ions, and each negative ion has a group of positive ions surrounding it. In most ionic solids, the ions are packed as tightly as possible.

26. Positive ions are always smaller than the atoms from which they are formed, because in forming the ion, the valence electron shell (or part of it) is "removed" from the atom.

27. In forming an anion, an atom gains additional electrons in its outermost (valence) shell. Having additional electrons in the valence shell increases the repulsive forces between electrons, and the outermost shell becomes larger to accommodate this.

28. Within a given horizontal row of the periodic chart, negative ions tend to be larger than positive ions because the negative ions contain a larger number of electrons in the valence shell. Within a vertical group of the periodic table, ionic size increases from top to bottom. In general, positive ions are smaller than the atoms they come from, whereas negative ions are larger than the atoms they come from.

     a. F^-
     b. Cl^-

Chemical Bonding

      c.   Ca

      d. I^-

29. Within a given horizontal row of the periodic chart, negative ions tend to be larger than positive ions because the negative ions contain a larger number of electrons in the valence shell. Within a vertical group of the periodic table, ionic size increases from top to bottom. In general, positive ions are smaller than the atoms they come from, whereas negative ions are larger than the atoms they come from.

      a. I^-

      b. Cl^-

      c. Cl^-

      d. S^{2-}

30. Valence electrons are those found in the outermost principal energy level of the atom. The valence electrons effectively represent the outside edge of the atom, and are the electrons most influenced by the electrons of another atom.

31. When atoms form covalent bonds, they try to attain a valence electronic configuration similar to that of the following noble gas element. When the elements in the first few horizontal rows of the periodic table form covalent bonds, they will attempt to gain configurations similar to the noble gases helium (2 valence electrons, duet rule), and neon and argon (8 valence electrons, octet rule).

32. When two atoms in a molecule are connected by a double bond, we are indicating that the atoms share two pairs of electrons (4 electrons) in completing their outermost shells. A simple molecule containing a double bond is ethene (ethylene),
C_2H_4 $H_2C::CH_2$

33.
    a.   C provides 4; each Br provides 7; total valence electrons = 32

    b.   N provides 5; each O provides 6; total valence electrons = 17

    c.   each C provides 4; each H provides 1; total valence electrons = 30

    d.   each O provides 6; each H provides 1; total valence electrons = 14

34.   a. NH_3 N provides 5 valence electrons.
                               Each H provides 1 valence electron.
                               Total valence electrons = 8

$$\text{H}-\overset{\cdot\cdot}{\underset{\underset{\text{H}}{|}}{\text{N}}}-\text{H}$$

b. CI₄

C provides 4 valence electrons.
Each I provides 7 valence electrons.
Total valence electrons = 32

```
        ..
      : I :
        |
  ..    |    ..
: I — C — I :
  ..    |    ..
        |
      : I :
        ..
```

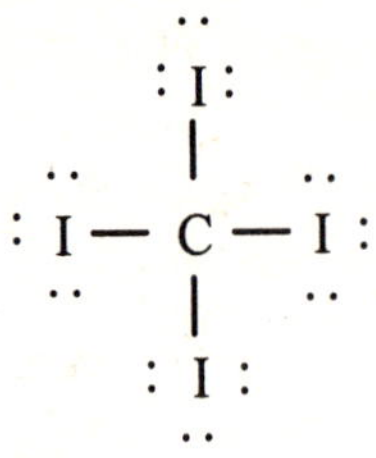

c. NCl₃

N provides 5 valence electrons.
Each Cl provides 7 valence electrons.
Total valence electrons = 26.

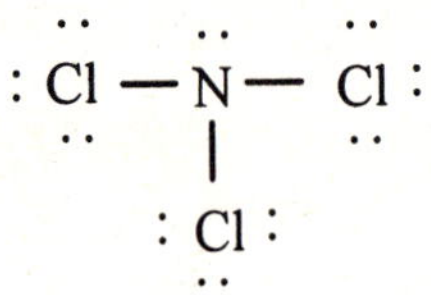

```
   ..    ..   ..
 : Cl — N — Cl :
   ..    |    ..
         |
       : Cl :
         ..
```

d. SiBr₄

Si provides 4 valence electrons.
Each Br provides 7 valence electrons.
Total valence electrons = 32

```
         ..
       : Br :
         |
   ..    |    ..
 : Br— Si — Br :
   ..    |    ..
         |
       : Br :
         ..
```

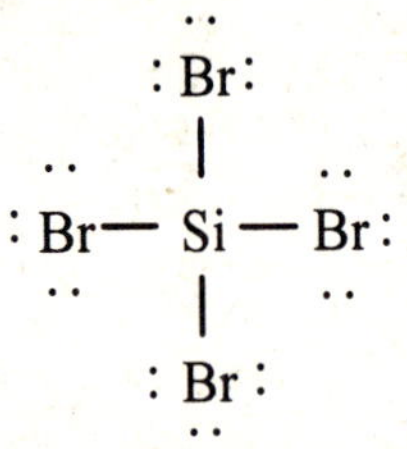

35. a. H₂S

Each H provides 1 valence electron.
S provides 6 valence electrons.
Total valence electrons = 8

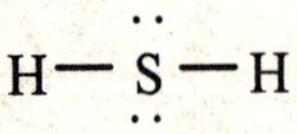

```
       ..
  H — S — H
       ..
```

b. SiF₄

Si provides 4 valence electrons.
Each F provides 7 valence electrons.
Total valence electrons = 32

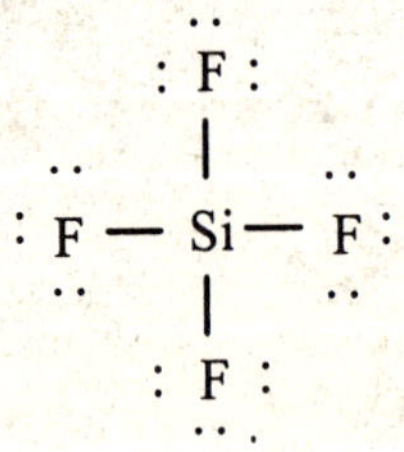

```
         ..
       : F :
         |
   ..    |    ..
 : F — Si — F :
   ..    |    ..
         |
       : F :
         ..
```

Chemical Bonding

c. C_2H_4 Each C provides 4 valence electrons.
Each H provides 1 valence electron.
Total valence electrons = 12

$$\begin{array}{c} H \\ \diagdown \\ C=C \\ \diagup \\ H \end{array} \begin{array}{c} H \\ \diagup \\ \\ \diagdown \\ H \end{array}$$

d. C_3H_8 Each C provides 4 valence electrons.
Each H provides 1 valence electron.
Total valence electrons = 20

$$\begin{array}{ccccc} & H & H & H & \\ & | & | & | & \\ H- & C- & C- & C- & H \\ & | & | & | & \\ & H & H & H & \end{array}$$

36. a. Cl_2O Each Cl provides 7 valence electrons.
O provides 6 valence electrons.
Total valence electrons = 20

$$:\!\ddot{C}l - \ddot{O} - \ddot{C}l\!:$$

b. CO_2 C provides 4 valence electrons.
Each O provides 6 valence electrons.
Total valence electrons = 16

$$\ddot{O}=C=\ddot{O} \longleftrightarrow :O\equiv C-\ddot{O}\!: \longleftrightarrow :\ddot{O}-C\equiv O:$$

c. SO_3 S provides 3 valence electrons.
Each O provides 3 valence electrons.
Total valence electrons = 24

$$\begin{array}{c}:\ddot{O}: \\ | \\ \phantom{:\ddot{O}}S\phantom{\ddot{O}:} \\ \diagup\diagdown \\ :\ddot{O} \ddot{O}:\end{array} \longleftrightarrow \begin{array}{c}:\ddot{O}: \\ \| \\ \phantom{:\ddot{O}}S\phantom{\ddot{O}:} \\ \diagup\diagdown \\ :\ddot{O} \ddot{O}:\end{array} \longleftrightarrow \begin{array}{c}:\ddot{O}: \\ | \\ \phantom{:\ddot{O}}S\phantom{\ddot{O}:} \\ \diagup\diagdown \\ :\ddot{O} \ddot{O}:\end{array}$$

37. a. SO_4^{2-} — S provides 6 valence electrons.
Each O provides 6 valence electrons.
The 2– charge means two additional valence electrons.
Total valence electrons = 32

$$\left[\begin{array}{c} \ddot{\underset{}{O}}\!: \\ | \\ :\!\ddot{O}\!-\!S\!-\!\ddot{O}\!: \\ | \\ :\!\ddot{\underset{}{O}}\!: \end{array}\right]^{2-}$$

b. PO_4^{3-} — P provides 5 valence electrons.
Each O provides 6 valence electrons.
The 3– charge means three additional valence electrons.
Total valence electrons = 32

$$\left[\begin{array}{c} \ddot{\underset{}{O}}\!: \\ | \\ :\!\ddot{O}\!-\!P\!-\!\ddot{O}\!: \\ | \\ :\!\ddot{\underset{}{O}}\!: \end{array}\right]^{3-}$$

c. SO_3^{2-} — S provides 6 valence electrons.
Each O provides 6 valence electrons.
The 2– charge means two additional valence electrons.
Total valence electrons = 26

$$\left[\begin{array}{c} \ddot{\underset{}{O}}\!: \\ | \\ :\!\ddot{O}\!-\!S\!-\!\ddot{O}\!: \end{array}\right]^{2-}$$

38. a. HPO_4^{2-} — H provides 1 valence electron.
P provides 5 valence electrons.
Each O provides 6 valence electrons
The 2– charge means two additional valence electrons.
Total valence electrons = 32

$$\left[\begin{array}{c} \ddot{\underset{}{O}}\!: \\ | \\ H\!-\!\ddot{O}\!-\!P\!-\!\ddot{O}\!: \\ | \\ :\!\ddot{\underset{}{O}}\!: \end{array}\right]^{2-}$$

b. $H_2PO_4^-$ Each H provides 1 valence electron.
P provides 5 valence electrons.
Each O provides 6 valence electrons.
The 1– charge means one additional valence electron.
Total valence electrons = 32

$$\left[\begin{array}{c} \ddot{\text{O}}: \\ | \\ \text{H}-\ddot{\text{O}}-\text{P}-\ddot{\text{O}}-\text{H} \\ | \\ :\ddot{\text{O}}: \end{array} \right]^{1-}$$

c. PO_4^{3-} P provides 5 valence electrons.
Each O provides 6 valence electrons.
The 3– charge means three additional valence electrons.
Total valence electrons = 32

$$\left[\begin{array}{c} :\ddot{\text{O}}: \\ | \\ :\ddot{\text{O}}-\text{P}-\ddot{\text{O}}: \\ | \\ :\ddot{\text{O}}: \end{array} \right]^{3-}$$

39. The geometric structure of the water molecule is bent (or V-shaped). There are four pairs of valence electrons on the oxygen atom of water (two pairs are bonding pairs, two pairs are non-bonding lone pairs). The H–O–H bond angle in water is approximately 106°.

40. The geometric structure of NH_3 is that of a trigonal pyramid. The nitrogen atom of NH_3 is surrounded by four electron pairs (three are bonding, one is a lone pair). The H–N–H bond angle is somewhat less than 109.5° (due to the presence of the lone pair).

41. BF_3 is described as having a trigonal planar geometric structure. The boron atom of BF_3 is surrounded by only three pairs of valence electrons. The F–B–F bond angle in BF_3 is 120°.

42. The geometric structure of CH_4 is that of a tetrahedron. The carbon atom of CH_4 is surrounded by four bonding electron pairs. The H–C–H bond angle is the characteristic angle of the tetrahedron, 109.5°.

43. The geometric structure of a molecule plays a very important part in its chemistry. For biological molecules, a slight change in the geometric structure of the molecule can completely destroy the molecule's usefulness to a cell, or can cause a destructive change in the cell.

44. The general molecular structure of a molecule is determined by *how many electron pairs* surround the central atom in the molecule, and by which of those electron pairs are used for *bonding* to the other atoms of the molecule. Nonbonding electron pairs on the central

atom do, however, cause minor changes in the bond angles, compared to the ideal regular geometric structure.

45. One of the valence electron pairs of the ammonia molecule is a *lone* pair with no hydrogen atom attached. This lone pair is part of the nitrogen atom and does not enter the description of the molecule's overall shape (aside from influencing the H–N–H bond angles between the remaining valence electron pairs).

46. In NF_3, the nitrogen atom has *four* pairs of valence electrons, whereas in BF_3, there are only *three* pairs of valence electrons around the boron atom. The nonbonding electron pair on nitrogen in NF_3 pushes the three F atoms out of the plane of the N atom.

47. If you draw the Lewis structures for these species, you will see that each of the indicated atoms is surrounded by *four pairs* of electrons with a *tetrahedral* orientation. Although the electron pairs are arranged tetrahedrally, realize that the overall geometric shape of the molecules may *not* be tetrahedral, depending on what atoms are *bonded* to the four pairs of electrons on the central atom.

48.
   a. tetrahedral (four electron pairs on C, and four atoms attached)
   b. nonlinear, V-shaped (four electron pairs on S, but only two atoms attached)
   c. tetrahedral (four electron pairs on Ge, and four atoms attached)

49.
   a. basically tetrahedral around the P atom (the hydrogen atoms are attached to two of the oxygen atoms and do not affect greatly the geometrical arrangement of the oxygen atoms around the phosphorus)
   b. tetrahedral (4 electron pairs on Cl, and 4 atoms attached)
   c. trigonal pyramidal (4 electron pairs on S, and 3 atoms attached)

50.
   a. <109.5°
   b. <109.5°
   c. 109.5°
   d. 109.5°

51. The CO_3^{2-} ion has a trigonal planar geometry, with bond angles of 120° degrees around the carbon atom. The molecule exhibits resonance.

52. Acetylene is a linear molecule with bond angles of 180°.

53. Resonance is said to exist when more than one valid Lewis structure can be drawn for a molecule. The actual bonding and structure of such molecules is thought to be somewhere "in between" the various possible Lewis structures. In such molecules, certain electrons are delocalized over more than one bond.

54. The bond with the larger electronegativity difference will be the more polar bond.
   a. S–F

Chemical Bonding

      b.   P–O

      c.   C–H

55. The bond energy of a chemical bond is the quantity of energy required to break the bond and separate the atoms.

56. In each case, the element *higher up* within a group of the periodic table has the higher electronegativity.

      a.   Be

      b.   N

      c.   F

57. In a diatomic molecule containing two different elements, the more electronegative atom will be the negative end of the molecule, and the *less* electronegative atom will be the positive end.

      a.   oxygen

      b.   bromine

      c.   iodine

58.
      a. Na^+ e. S^{2-}

      b. I^- f. Mg^{2+}

      c. K^+ g. Al^{3+}

      d. Ca^{2+} h. N^{3-}

59.
      a.   nonlinear (V-shaped)

      b.   trigonal planar

      c.   basically trigonal planar around the C (the H is attached to one of the O atoms, and distorts the shape around the carbon only slightly)

      d.   linear

60. In a covalent bond between two atoms of the same element, the electron pair is shared equally and the bond is nonpolar; with a bond between atoms of different elements, the electron pair is unequally shared and the bond is polar (assuming the elements have different electronegativities).

Chapter 13 Gases

1. Solids and liquids have essentially fixed volumes and are not able to be compressed easily. Gases have volumes that depend on their conditions, and can be compressed or expanded by changes in those conditions. Although the particles of matter in solids are essentially fixed in position (the solid is rigid), the particles in liquids and gases are free to move.

2. The "pressure of the atmosphere" represents the weight of the several- mile-thick layer of gases pressing down on every surface of the earth. Pressure, in general, represents a force exerted over a particular area, and the pressure of the atmosphere corresponds to a pressure of nearly 15 pounds per square inch on the surface of the earth.

3. A simple mercury barometer is a tube filled with mercury inverted over a reservoir containing mercury which is open to the atmosphere. When the tube is inverted, the mercury falls to a level at which the pressure of the atmosphere is sufficient to support the column of mercury. One standard atmosphere of pressure is taken to be the pressure capable of supporting a column of mercury to a height of 760.0 mm above the reservoir level.

4.
   a. $105.2 \text{ kPa} \times \dfrac{1 \text{ atm}}{101.325 \text{ kPa}} = 1.038 \text{ atm}$

   b. $75.2 \text{ cm Hg} \times \dfrac{10 \text{ mm Hg}}{1 \text{ cm Hg}} \times \dfrac{1 \text{ atm}}{760 \text{ mm Hg}} = 0.989 \text{ atm}$

   c. $752 \text{ mm Hg} \times \dfrac{1 \text{ atm}}{760 \text{ mm Hg}} = 0.989 \text{ atm}$

   d. 767 torr = 767 mm Hg

   $767 \text{ torr} \times \dfrac{1 \text{ atm}}{760 \text{ torr}} = 1.01 \text{ atm}$

5.
   a. $0.9975 \text{ atm} \times \dfrac{760 \text{ mm Hg}}{1 \text{ atm}} = 758.1 \text{ mm Hg}$

   b. $225{,}400 \text{ Pa} \times \dfrac{760 \text{ mm Hg}}{101{,}325 \text{ Pa}} = 1691 \text{ mm Hg}$

   c. $99.7 \text{ kPa} \times \dfrac{760 \text{ mm Hg}}{101.325 \text{ kPa}} = 748 \text{ mm Hg}$

   d. $1.078 \text{ atm} \times \dfrac{760 \text{ mm Hg}}{1 \text{ atm}} = 819.3 \text{ mm Hg}$

6.
   a. $774 \text{ torr} \times \dfrac{101{,}325 \text{ Pa}}{760 \text{ torr}} = 1.03 \times 10^5 \text{ Pa}$

   b. $0.965 \text{ atm} \times \dfrac{101{,}325 \text{ Pa}}{1 \text{ atm}} = 9.78 \times 10^4 \text{ Pa}$

Gases

    c.   112.5 kPa = 1.125×10^5 Pa

    d.   801 mm Hg $\times \dfrac{101{,}325 \text{ Pa}}{760 \text{ mm Hg}} = 1.07 \times 10^5$ Pa

7.   a. P_1 = 785 mm Hg P_2 = 700 mm Hg

           V_1 = 53.2 mL V_2 = ? mL

$$V_2 = \dfrac{P_1 V_1}{P_2} = \dfrac{(53.2 \text{ mL})(785 \text{ mm Hg})}{700 \text{ mm Hg}} = 59.7 \text{ mL}$$

    b. P_1 = 1.67 atm P_2 = ? atm

           V_1 = 2.25 L V_2 = 2.00 L

$$P_2 = \dfrac{P_1 V_1}{V_2} = \dfrac{(1.67 \text{ atm})(2.25 \text{ L})}{2.00 \text{ L}} = 1.88 \text{ atm}$$

    c. P_1 = 695 mm Hg P_2 = 1.51 atm = 1148 mm Hg

           V_1 = 5.62 L V_2 = ? L

$$V_2 = \dfrac{P_1 V_1}{P_2} = \dfrac{(695 \text{ mm Hg})(5.62 \text{ L})}{1148 \text{ mm Hg}} = 3.40 \text{ L}$$

8.   a. P_1 = 1.07 atm P_2 = 2.14 atm

           V_1 = 291 mL V_2 = ? mL

$$V_2 = \dfrac{P_1 V_1}{P_2} = \dfrac{(1.07 \text{ atm})(291 \text{ mL})}{2.14 \text{ atm}} = 146 \text{ mL}$$

    b. P_1 = 755 mm Hg P_2 = 3.51 atm = 2668 mm Hg

           V_1 = 1.25 L V_2 = ? L

$$V_2 = \dfrac{P_1 V_1}{P_2} = \dfrac{(755 \text{ mm Hg})(1.25 \text{ L})}{2668 \text{ mm Hg}} = 0.354 \text{ L}$$

    c. P_1 = 101.4 kPa = 760.6 mm Hg P_2 = ? mm Hg

           V_1 = 2.71 L V_2 = 3.00 L

$$P_2 = \dfrac{P_1 V_1}{V_2} = \dfrac{(760.6 \text{ mm Hg})(2.71 \text{ L})}{(3.00 \text{ L})} = 687 \text{ mm Hg}$$

9.   If the pressure exerted on the gas in the balloon is decreased, the volume of the gas in the balloon will increase in inverse proportion to the factor by which the pressure was changed. The factor in this example is (1.01 atm/0.562 atm) = 1.80

10. P_1 = 760 mm = 1.00 atm P_2 = ? atm

           V_1 = 1.00 L V_2 = 50.0 mL = 0.0500 L

$$P_2 = \frac{P_1V_1}{V_2} = \frac{(1.00 \text{ atm})(1.00 \text{ L})}{(0.0500 \text{ L})} = 20.0 \text{ atm}$$

11. Absolute zero is the lowest temperature than can exist. Absolute zero is the temperature at which the volume of an ideal gas sample would be predicted to become zero. Absolute zero is the zero-point on the Kelvin temperature scale (and corresponds to –273°C).

12. Charles's law states that the volume of an ideal gas sample varies linearly with the absolute temperature of the gas sample. In an experiment performed to determine absolute zero, the volume of a sample of gas is measured at several convenient temperatures (e.g., between 0° and 100°C) and the data is plotted. The straight line obtained is then extrapolated to the point where the volume of the gas would become zero. The temperature at which the volume of the gas would be predicted to become zero is then absolute zero.

13. $V_1 = 45.0$ mL $\qquad V_2 = ?$ mL
    $T_1 = 26.5°C = 300$ K $\qquad T_2 = 55.2°C = 328$ K

    $$V_2 = \frac{V_1T_2}{T_1} = \frac{(45.0 \text{ mL})(328 \text{ K})}{(300 \text{ K})} = 49.2 \text{ mL}$$

14. a. $V_1 = 25.0$ L $\qquad V_2 = 50.0$ L
    $T_1 = 0°C = 273$ K $\qquad T_2 = ?$ °C

    $$T_2 = \frac{V_2T_1}{V_1} = \frac{(50.0 \text{ L})(273 \text{ K})}{(25.0 \text{ L})} = 546 \text{ K} = 273°C$$

    b. $V_1 = 247$ mL $\qquad V_2 = 255$ mL
    $T_1 = 25°C = 298$ K $\qquad T_2 = ?$ °C

    $$T_2 = \frac{V_2T_1}{V_1} = \frac{(255 \text{ mL})(298 \text{ K})}{(247 \text{ mL})} = 308 \text{ K} = 35°C$$

    c. $V_1 = 1.00$ mL $\qquad V_2 = ?$ mL
    $T_1 = -272°C = 1$ K $\qquad T_2 = 25°C = 298$ K

    $$V_2 = \frac{V_1T_2}{T_1} = \frac{(1.00 \text{ mL})(298 \text{ K})}{(1 \text{ K})} = 298 \text{ mL}$$

15. a. $V_1 = 2.01 \times 10^2$ L $\qquad V_2 = 5.00$ L
    $T_1 = 1150°C = 1423$ K $\qquad T_2 = ?$ °C

    $$T_2 = \frac{V_2T_1}{V_1} = \frac{(5.00 \text{ L})(1423 \text{ K})}{(2.01 \times 10^2 \text{ L})} = 35.4 \text{ K} = -238°C$$

    b. $V_1 = 44.2$ mL $\qquad V_2 = ?$ mL
    $T_1 = 298$ K $\qquad T_2 = 0$ K

Gases

$$V_2 = \frac{V_1 T_2}{T_1} = \frac{(44.2 \text{ mL})(0 \text{ K})}{(298 \text{ K})} = 0 \text{ mL (0 K is absolute zero)}$$

c. $V_1 = 44.2$ mL $V_2 = ?$ mL

     $T_1 = 298$ K $T_2 = 0°C = 273$ K

$$V_2 = \frac{V_1 T_2}{T_1} = \frac{(44.2 \text{ mL})(273 \text{ K})}{(298 \text{ K})} = 40.5 \text{ mL}$$

16. $24°C = 297$ K $-272°C = 1$ K

$$5.00 \text{ L} \times \frac{1 \text{ K}}{297 \text{ K}} = 0.0168 \text{ L} = 0.02 \text{ L}$$

17. You should be able to answer these without having to set up a formal calculation. Charles's law says that the volume of a gas sample is *directly* proportional to its absolute temperature. So if a sample of neon has a volume of 266 mL at 25.2°C (298 K), then the volume will become half as big at half the absolute temperature (149 K, –124°C). The volume of the gas sample will become twice as big at twice the absolute temperature (596 K, 323°C).

18. $25°C + 273 = 298$ K $54°C + 273 = 327$ K

$$500. \text{ mL} \times \frac{327 \text{ K}}{298 \text{ K}} = 549 \text{ mL}$$

19. $V_1 = 652$ mL $V_2 = ?$ L

     $n_1 = 0.214$ mol $n_2 = 0.375$ mol

$$652 \text{ mL} \times \frac{0.375 \text{ mol}}{0.214 \text{ mol}} = 1143 \text{ mL} = 1.14 \text{ L}$$

20. $V_1 = 100.$ L $V_2 = ?$ L

     $n_1 = 46.2$ g/32.00 g mol^{-1} $n_2 = 5.00$ g/32.00 g mol^{-1}

$$V_2 = \frac{V_1 n_2}{n_1} = \frac{(100. \text{ L})(5.00 \text{ g}/32.00 \text{ g mol}^{-1})}{(46.2 \text{ g}/32.00 \text{ g mol}^{-1})} = 10.8 \text{ L}$$

Note that the *molar mass* of the O_2 gas cancels out in this calculation. Since the number of moles of Ne (or any gas) present in a sample is *directly proportional* to the mass of the gas sample, the problem could also have been set up directly in terms of the masses.

21. Real gases most closely approach ideal gas behavior under conditions of relatively high temperatures (0°C or higher) and relatively low pressures (1 atm or lower).

22. For an ideal gas, $PV = nRT$ is true under any conditions. Consider a particular sample of gas (so that n remains constant) at a particular fixed temperature (so that T remains constant also). Suppose that at pressure P_1 the volume of the gas sample is V_1. Then for

Copyright © Houghton Mifflin Company. All rights reserved.

this set of conditions, the ideal gas equation would be given by

$$P_1V_1 = nRT$$

If we then change the pressure of the gas sample to a new pressure P_2, the volume of the gas sample changes to a new volume V_2. For this new set of conditions, the ideal gas equation would be given by

$$P_2V_2 = nRT$$

Since the right-hand sides of these equations are equal to the same quantity (since we defined n and T to be constant), then the left-hand sides of the equations must also be equal, and we obtain the usual form of Boyle's law.

$$P_1V_1 = P_2V_2$$

23.   a. $P = 782$ mm Hg = 1.03 atm

           $T = 27°C = 300$ K

$$V = \frac{nRT}{P} = \frac{(0.210 \text{ mol})(0.08206 \text{ L atm mol}^{-1} \text{ K}^{-1})(300. \text{ K})}{(1.03 \text{ atm})} = 5.02 \text{ L}$$

      b. $V = 644$ mL = 0.644 L

$$P = \frac{nRT}{V} = \frac{(0.0921 \text{ mol})(0.08206 \text{ L atm mol}^{-1} \text{ K}^{-1})(303 \text{ K})}{(0.644 \text{ L})} = 3.56 \text{ atm}$$

           $P = 3.56$ atm $= 2.70 \times 10^3$ mm Hg

      c. $P = 745$ mm = 0.980 atm

$$T = \frac{PV}{nR} = \frac{(0.980 \text{ atm})(11.2 \text{ L})}{(0.401 \text{ mol})(0.08206 \text{ L atm mol}^{-1} \text{ K}^{-1})} = 334 \text{ K}$$

24.   a. $T = 25°C = 298$ K

$$V = \frac{(0.00831 \text{ mol})(0.08206 \text{ L atm mol}^{-1} \text{ K}^{-1})(298 \text{ K})}{(1.01 \text{ atm})} = 0.201 \text{ L}$$

      b. $V = 602$ mL = 0.602 L

$$P = \frac{(8.01 \times 10^{-3} \text{ mol})(0.08206 \text{ L atm mol}^{-1} \text{ K}^{-1})(310 \text{ K})}{(0.602 \text{ L})} = 0.338 \text{ atm}$$

      c. $V = 629$ mL = 0.629 L

           $T = 35°C = 308$ K

$$n = \frac{(0.998 \text{ atm})(0.629 \text{ L})}{(0.08206 \text{ L atm mol}^{-1} \text{ K}^{-1})(308 \text{ K})} = 2.48 \times 10^{-2} \text{ mol}$$

Gases 133

25. molar mass of N_2 = 28.02 g   58.2°C = 331 K

$n = 4.24 \text{ g } N_2 \times \dfrac{1 \text{ mol } N_2}{28.02 \text{ g } N_2} = 0.151 \text{ mol } N_2$

$V = nRT/P = \dfrac{(0.151 \text{ mol})(0.08206 \text{ L atm mol}^{-1} \text{ K}^{-1})(331 \text{ K})}{(2.04 \text{ atm})} = 2.01 \text{ L}$

26. The number of moles of *any* ideal gas that can be contained in the tank under the given conditions can first be calculated.

$T = 24°C = 297 \text{ K}$

$n = \dfrac{PV}{RT} = \dfrac{(135 \text{ atm})(200 \text{ L})}{(0.08206 \text{ L atm mol}^{-1} \text{ K}^{-1})(297 \text{ K})} = 1.11 \times 10^3 \text{ mol gas}$

Molar masses: He, 4.003 g; H_2, 2.016 g

for He: $1.11 \times 10^3 \text{ mol He} \times \dfrac{4.003 \text{ g}}{1 \text{ mol}} = 4.44 \times 10^3 \text{ g He} = 4.44 \text{ kg He}$

for H_2: $1.11 \times 10^3 \text{ mol } H_2 \times \dfrac{2.016 \text{ g}}{1 \text{ mol}} = 2.24 \times 10^3 \text{ g } H_2 = 2.24 \text{ kg } H_2$

27. Molar mass of N_2 = 28.02 g

$16.3 \text{ g } N_2 \times \dfrac{1 \text{ mol}}{28.02 \text{ g}} = 0.582 \text{ mol } N_2$

$T = \dfrac{PV}{nR} = \dfrac{(1.25 \text{ atm})(25.0 \text{ L})}{(0.582 \text{ mol})(0.08206 \text{ L atm mol}^{-1} \text{ K}^{-1})} = 654 \text{ K} = 381°C$

28. Molar mass of O_2 = 32.00 g   56.2 kg = 5.62×10^4 g

$5.62 \times 10^4 \text{ g} \times \dfrac{1 \text{ mol}}{32.00 \text{ g}} = 1.76 \times 10^3 \text{ mol}$

$T = 21°C = 294 \text{ K}$

$P = \dfrac{nRT}{V} = \dfrac{(1.76 \times 10^3 \text{ mol})(0.08206 \text{ L atm mol}^{-1} \text{ K}^{-1})(294 \text{ K})}{(125 \text{ L})}$

$P = 340 \text{ atm}$

29. $P_1 = 0.981 \text{ atm}$ $P_2 = 1.15 \text{ atm}$

$V_1 = 125 \text{ mL}$ $V_2 = ? \text{ mL}$

$T_1 = 100°C = 373 \text{ K}$ $T_2 = 25°C = 298 \text{ K}$

$V_2 = \dfrac{P_1 V_1 T_2}{P_2 T_1} = \dfrac{(0.981 \text{ atm})(125 \text{ mL})(298 \text{ K})}{(1.15 \text{ atm})(373 \text{ K})} = 85.2 \text{ mL}$

30. $P_1 = 1.05 \text{ atm}$ $P_2 = 0.997 \text{ atm}$

$V_1 = 459 \text{ mL}$ $V_2 = ? \text{ mL}$

$T_1 = 27°C = 300.\ K \qquad T_2 = 15°C = 288\ K$

$$V_2 = \frac{P_1 V_1 T_2}{P_2 T_1} = \frac{(1.05\ atm)(459\ mL)(288\ K)}{(0.997\ atm)(300\ K)} = 464\ mL$$

31. In deriving the ideal gas law, we assume that the molecules of gas themselves occupy no volume, and that the molecules do not interact with each other. Under these conditions, there is no difference between gas molecules of different substances (other than their masses) as far as the bulk behavior of the gas is concerned. Each gas behaves independently of other gases present, and the overall properties of the sample are determined by the overall quantity of gas present.

    $P_{total} = P_1 + P_2 + ...\ P_n$ where n is the number of individual gases present in the mixture

32. As a gas is bubbled through water, the bubbles of gas become saturated with water vapor, thus forming a gaseous mixture. The total pressure in a sample of gas which has been collected by bubbling through water is made up of two components: the pressure of the gas of interest and the pressure of water vapor. The partial pressure of the gas of interest is then the total pressure of the sample minus the vapor pressure of water.

33. molar masses: O_2, 32.00 g; He, 4.003 g $\qquad 65°C + 273 = 338\ K$

    $4.0\ g\ O_2 \times \dfrac{1\ mol\ O_2}{32.0\ g\ O_2} = 0.125\ mol\ O_2$

    $4.0\ g\ He \times \dfrac{1\ mol\ He}{4.003\ g\ He} = 0.999\ mol\ He$

    $P_{oxygen} = n_{oxygen} RT/V = \dfrac{(0.125\ mol)(0.08206\ L\ atm\ mol^{-1}\ K^{-1})(338\ K)}{(5.0\ L)}$

    $P_{oxygen} = 0.693\ atm = 0.69\ atm$

    $P_{helium} = n_{helium} RT/V = \dfrac{(0.999\ mol)(0.08206\ L\ atm\ mol^{-1}\ K^{-1})(338\ K)}{(5.0\ L)}$

    $P_{helium} = 5.54\ atm = 5.5\ atm$

    $P_{total} = 0.693\ atm + 5.54\ atm = 6.233\ atm = 6.2\ atm$

34. Total moles of gas = 3.0 mol + 2.0 mol + 1.0 mol = 6.0 mol

    $P_{nitrogen} = 10.0\ atm \times \dfrac{3.0\ mol}{6.0\ mol} = 5.0\ atm$

    $P_{oxygen} = 10.0\ atm \times \dfrac{2.0\ mol}{6.0\ mol} = 3.3\ atm$

    $P_{carbon\ dioxide} = 10.0\ atm \times \dfrac{1.0\ mol}{6.0\ mol} = 1.7\ atm$

35. $P_{oxygen} = P_{total} - P_{water\ vapor} = 772 - 26.7 = 745\ torr$

Gases

36. $P_{oxygen} = P_{total} - P_{water\ vapor} = 755 - 23 = 732$ mm Hg $= 0.9632$ atm

    $T = 24°C + 273 = 297$ K

    $V = 500.$ mL $= 0.500$ L

    $n = PV/RT = \dfrac{(0.9632\ atm)(0.500\ L)}{(0.08206\ L\ atm\ mol^{-1}\ K^{-1})(297\ K)} = 1.98 \times 10^{-2}$ mol O_2

37. A *law* is a statement that precisely expresses generally observed behavior. A *theory* consists of a set of assumptions/hypotheses that is put forth to *explain* the observed behavior of matter. Theories attempt to explain natural laws.

38. A theory is successful if it explains known experimental observations. Theories which have been successful in the past may not be successful in the future (for example, as technology evolves, more sophisticated experiments may be possible in the future).

39. We assume that the volume of the molecules themselves in a gas sample is negligible compared to the bulk volume of the gas sample: this helps us to explain why gases are so compressible.

40. Chemists believe the pressure exerted by a gas sample on the walls of its container arises from collisions between the gas molecules and the walls of the container.

41. kinetic energy

42. The temperature of a gas reflects, on average, how rapidly the molecules in the gas are moving. At high temperatures, the particles are moving very fast and collide with the walls of the container frequently, whereas at low temperatures, the molecules are moving more slowly and collide with the walls of the container infrequently. The Kelvin temperature is directly proportional to the average kinetic energy of the particles in a gas.

43. If the temperature of a sample of gas is increased, the average kinetic energy of the particles of gas increases. This means that the speeds of the particles increase. If the particles have a higher speed, they will hit the walls of the container more frequently and with greater force, thereby increasing the pressure.

44. Any gas that does not follow the ideal gas law is not behaving ideally. In addition, if there are reasons to believe that the assumptions of the kinetic molecular theory are poor assumptions, the gas sample is not behaving ideally. The fact that gases condense into liquids, for example, shows non-ideal behavior.

45. When the volume of a gas is decreased, the gas particles take up a greater percentage of the volume of the container. The assumption in the KMT that gas particles take up a negligible volume is less correct.

46. $CaCO_3(s) \rightarrow CO_2(g) + CaO(s)$

    molar mass $CaCO_3 = 100.1$ g

    15.2 g $CaCO_3 \times \dfrac{1\ mol\ CaCO_3}{100.1\ g\ CaCO_3} = 0.152$ mol $CaCO_3$

From the balanced chemical equation, if 0.152 mol $CaCO_3$ reacts, 0.152 mol of CO_2 will result.

STP: 1.00 atm, 273 K

$$V = nRT/P = \frac{(0.152 \text{ mol})(0.08206 \text{ L atm mol}^{-1} \text{ K}^{-1})(273 \text{ K})}{(1.00 \text{ atm})} = 3.41 \text{ L}$$

47. $C_3H_8(g) + 5O_2(g) \rightarrow 3CO_2(g) + 4H_2O(g)$

    25°C + 273 = 298 K

    molar mass C_3H_8 = 44.09 g

    $$5.53 \text{ g } C_3H_8 \times \frac{1.0 \text{ mol } C_3H_8}{44.09 \text{ g } C_3H_8} = 0.1254 \text{ mol } C_3H_8$$

    $$0.1254 \text{ mol } C_3H_8 \times \frac{5 \text{ mol } O_2}{1 \text{ mol } C_3H_8} = 0.6270 \text{ mol } O_2$$

    $$V = nRT/P = \frac{(0.6270 \text{ mol } O_2)(0.08206 \text{ L atm mol}^{-1} \text{ K}^{-1})(298 \text{ K})}{(1.04 \text{ atm})} = 14.7 \text{ L } O_2$$

48. 27°C = 300 K   26°C = 299 K

    $$\text{mol } NH_3 \text{ present} = \frac{(1.02 \text{ atm})(4.21 \text{ L})}{(0.08206 \text{ L atm mol}^{-1} \text{ K}^{-1})(300 \text{ K})} = 0.174 \text{ mol } NH_3$$

    $$\text{mol HCl present} = \frac{(0.998 \text{ atm})(5.35 \text{ L})}{(0.08206 \text{ L atm mol}^{-1} \text{ K}^{-1})(299 \text{ K})} = 0.218 \text{ mol HCl}$$

    NH_3 and HCl react on a 1:1 basis: NH_3 is the limiting reactant.

    molar mass NH_4Cl = 53.49 g

    $$0.174 \text{ mol } NH_3 \times \frac{1 \text{ mol } NH_4Cl}{1 \text{ mol } NH_3} \times \frac{53.49 \text{ g } NH_4Cl}{1 \text{ mol } NH_4Cl} = 9.31 \text{ g } NH_4Cl \text{ produced}$$

49. Molar mass of Mg_3N_2 = 100.95 g

    $$10.3 \text{ g } Mg_3N_2 \times \frac{1 \text{ mol}}{100.95 \text{ g}} = 0.102 \text{ mol } Mg_3N_2$$

    From the balanced chemical equation, the amount of NH_3 produced will be

    $$0.102 \text{ mol } Mg_3N_2 \times \frac{2 \text{ mol } NH_3}{1 \text{ mol } Mg_3N_2} = 0.204 \text{ mol } NH_3$$

    T = 24°C = 297 K P = 752 mm Hg = 0.989 atm

    $$V = \frac{nRT}{P} = \frac{(0.204 \text{ mol})(0.08206 \text{ L atm mol}^{-1} \text{ K}^{-1})(297 \text{ K})}{(0.989 \text{ atm})} = 5.03 \text{ L}$$

    This assumes that the ammonia was collected dry.

50. Molar masses: He, 4.003 g; H_2, 2.016 g

$$14.2 \text{ g He} \times \frac{1 \text{ mol He}}{4.003 \text{ g He}} = 3.55 \text{ mol He}$$

$$21.6 \text{ g } H_2 \times \frac{1 \text{ mol } H_2}{2.016 \text{ g } H_2} = 10.7 \text{ mol}$$

total moles = 3.55 mol + 10.7 mol = 14.3 mol

28°C = 301 K

$$V = \frac{nRT}{P} = \frac{(14.3 \text{ mol})(0.08206 \text{ L atm mol}^{-1} \text{ K}^{-1})(301 \text{ K})}{(0.985 \text{ atm})} = 359 \text{ L}$$

51. $P_1 = 1.47$ atm $P_2 = 1.00$ atm (Standard Pressure)
    $V_1 = 145$ mL $V_2 = ?$ mL
    $T_1 = 44°C = 317$ K $T_2 = 0°C = 273$ K (Standard Temperature)

$$V_2 = \frac{P_1 V_1 T_2}{P_2 T_1} = \frac{(1.47 \text{ atm})(145 \text{ mL})(273 \text{ K})}{(1.00 \text{ atm})(317 \text{ K})} = 184 \text{ mL}$$

52. molar masses: O_2, 32.00 g; N_2, 28.02 g; CO_2, 44.01 g; Ne, 20.18 g

$$5.00 \text{ g } O_2 \times \frac{1 \text{ mol } O_2}{32.00 \text{ g } O_2} = 0.1563 \text{ mol } O_2$$

$$5.00 \text{ g } N_2 \times \frac{1 \text{ mol } N_2}{28.02 \text{ g } N_2} = 0.1784 \text{ mol } N_2$$

$$5.00 \text{ g } CO_2 \times \frac{1 \text{ mol } CO_2}{44.01 \text{ g } CO_2} = 0.1136 \text{ mol } CO_2$$

$$5.00 \text{ g Ne} \times \frac{1 \text{ mol Ne}}{20.18 \text{ g Ne}} = 0.2478 \text{ mol Ne}$$

Total moles of gas = 0.1563 + 0.1784 + 0.1136 + 0.2478 = 0.6961 mol

22.4 L is the volume occupied by one mole of any ideal gas at STP. This would apply even if the gas sample is a *mixture* of individual gases.

$$0.6961 \text{ mol} \times \frac{22.4 \text{ L}}{1 \text{ mol}} = 15.59 \text{ L} = 15.6 \text{ L}$$

The *partial pressure* of each individual gas in the mixture will be related to what *fraction* on a mole basis each gas represents in the mixture.

$$P_{\text{oxygen}} = 1.00 \text{ atm} \times \frac{0.1563 \text{ mol } O_2}{0.6961 \text{ mol total}} = 0.225 \text{ atm } O_2$$

$$P_{\text{nitrogen}} = 1.00 \text{ atm} \times \frac{0.1784 \text{ mol } N_2}{0.6961 \text{ mol total}} = 0.256 \text{ atm } N_2$$

$$P_{\text{carbon dioxide}} = 1.00 \text{ atm} \times \frac{0.1136 \text{ mol CO}_2}{0.6961 \text{ mol total}} = 0.163 \text{ atm CO}_2$$

$$P_{\text{neon}} = 1.00 \text{ atm} \times \frac{0.2478 \text{ mol Ne}}{0.6961 \text{ mol total}} = 0.356 \text{ atm Ne}$$

53. $2\text{Na}(s) + \text{Cl}_2(g) \rightarrow 2\text{NaCl}(s)$

    molar mass Na = 22.99 g

    $$4.81 \times \frac{1 \text{ mol Na}}{22.99 \text{ g Na}} = 0.2092 \text{ mol Na}$$

    $$0.2092 \text{ mol Na} \times \frac{1 \text{ mol Cl}_2}{2 \text{ mol Na}} = 0.1046 \text{ mol Cl}_2$$

    $$0.1046 \text{ mol Cl}_2 \times \frac{22.4 \text{ L}}{1 \text{ mol}} = 2.34 \text{ L Cl}_2 \text{ at STP}$$

54. $2\text{K}_2\text{MnO}_4(aq) + \text{Cl}_2(g) \rightarrow 2\text{KMnO}_4(s) + 2\text{KCl}(aq)$

    molar mass KMnO$_4$ = 158.0 g

    $$10.0 \text{ g KMnO}_4 \times \frac{1 \text{ mol KMnO}_4}{158.0 \text{ g KMnO}_4} = 0.06329 \text{ mol KMnO}_4$$

    $$0.06329 \text{ mol KMnO}_4 \times \frac{1 \text{ mol Cl}_2}{2 \text{ mol KMnO}_4} = 0.03165 \text{ mol Cl}_2$$

    $$0.03165 \text{ mol Cl}_2 \times \frac{22.4 \text{ L}}{1 \text{ mol}} = 0.709 \text{ L} = 709 \text{ mL}$$

55. First determine what volume the helium in the tank would have if it were at a pressure of 755 mm Hg (corresponding to the pressure the gas will have in the balloons).

    8.40 atm = 6384 mm Hg

    $$V_2 = (25.2 \text{ L}) \times \frac{6384 \text{ mm Hg}}{755 \text{ mm Hg}} = 213$$

    Allowing for the fact that 25.2 L of He will have to remain in the tank, this leaves 213 − 25.2 = 187.8 L of He for filling the balloons.

    $$187.8 \text{ L He} \times \frac{1 \text{ balloon}}{1.50 \text{ L He}} = 125 \text{ balloons}$$

56. A decrease in temperature would tend to make the volume of the weather balloon *decrease*. Since the overall volume of a weather balloon *increases* when it rises to higher altitudes, the contribution to the new volume of the gas from the decrease in pressure must be more important than the decrease in temperature (the temperature change in kelvins is not as dramatic as it seems in degrees Celsius).

57. $2\text{S}(s) + 3\text{O}_2(g) \rightarrow 2\text{SO}_3(g)$

    350.°C + 273 = 623 K

Gases

molar mass S = 32.07 g

$$5.00 \text{ g} \times \frac{1 \text{ mol S}}{32.07 \text{ g S}} = 0.1559 \text{ mol S}$$

$$0.1559 \text{ mol} \times \frac{3 \text{ mol O}_2}{2 \text{ mol S}} = 0.2339 \text{ mol O}_2$$

$$V = nRT/P = \frac{(0.2339 \text{ mol})(0.08206 \text{ L atm mol}^{-1} \text{ K}^{-1})(623 \text{ K})}{(5.25 \text{ atm})} = 2.28 \text{ L O}_2$$

58. Assume the pressure at sea level to be 1 atm (760 mm Hg). Calculate the volume the balloon would have if it rose to the point where the pressure has dropped to 500 mm Hg. If this calculated volume is greater than the balloon's specified maximum volume (2.5 L) the balloon will burst.

$$2.0 \text{ L} \times \frac{760 \text{ mm Hg}}{500 \text{ mm Hg}} = 3.0 \text{ L} > 2.5 \text{ L. The balloon will burst.}$$

59. 22°C + 273 = 295 K   100°C + 273 = 373 K

$$729 \text{ mL} \times \frac{373 \text{ K}}{295 \text{ K}} = 922 \text{ mL}$$

60. $P_1 = 1.0$ atm $P_2 = 220$ torr $= 0.289$ atm
    $V_1 = 1.0$ L $V_2 = ?$
    $T_1 = 23°C + 273 = 296$ K $T_2 = -31°C = 242$ K

$$V_2 = \frac{T_2 P_1 V_1}{T_1 P_2} = \frac{(242 \text{ K})(1.0 \text{ atm})(1.0 \text{ L})}{(295 \text{ K})(0.289 \text{ atm})} = 2.8 \text{ L}$$

61. $2Cu_2S(s) + 3O_2(g) \rightarrow 2Cu_2O(s) + 2SO_2(g)$

molar mass Cu_2S = 159.2 g

$$25 \text{ g Cu}_2\text{S} \times \frac{1 \text{ mol Cu}_2\text{S}}{159.2 \text{ g Cu}_2\text{S}} = 0.1570 \text{ mol Cu}_2\text{S}$$

$$0.1570 \text{ mol Cu}_2\text{S} \times \frac{3 \text{ mol O}_2}{2 \text{ mol Cu}_2\text{S}} = 0.2355 \text{ mol O}_2$$

27.5°C + 273 = 301 K

$$V_{\text{oxygen}} = \frac{(0.2355 \text{ mol})(0.08206 \text{ L atm mol}^{-1} \text{ K}^{-1})(301 \text{ K})}{(0.998 \text{ atm})} = 5.8 \text{ L O}_2$$

$$0.1570 \text{ mol Cu}_2\text{S} \times \frac{2 \text{ mol SO}_2}{2 \text{ mol Cu}_2\text{S}} = 0.1570 \text{ mol SO}_2$$

$$V_{\text{sulfur dioxide}} = \frac{(0.1570 \text{ mol})(0.08206 \text{ L atm mol}^{-1} \text{ K}^{-1})(301 \text{ K})}{(0.998 \text{ atm})}$$

$V_{\text{sulfur dioxide}} = 3.9 \text{ L } SO_2$

62. molar masses: He, 4.003 g; Ar, 39.95 g; Ne, 20.18 g

    $5.0 \text{ g He} \times \dfrac{1 \text{ mol He}}{4.003 \text{ g He}} = 1.249 \text{ mol He}$

    $1.0 \text{ g Ar} \times \dfrac{1 \text{ mol Ar}}{39.95 \text{ g Ar}} = 0.02503 \text{ mol Ar}$

    $3.5 \text{ g Ne} \times \dfrac{1 \text{ mol Ne}}{20.18 \text{ g Ne}} = 0.1734 \text{ mol Ne}$

    Total moles of gas = 1.249 + 0.02503 + 0.1734 = 1.447 mol

    22.4 L is the volume occupied by one mole of any ideal gas at STP. This would apply even if the gas sample is a *mixture* of individual gases.

    $1.447 \text{ mol} \times \dfrac{22.4 \text{ L}}{1 \text{ mol}} = 32 \text{ L total volume for the mixture}$

    The *partial pressure* of each individual gas in the mixture will be related to what *fraction* on a mole basis each gas represents in the mixture.

    $P_{He} = 1.00 \text{ atm} \times \dfrac{1.249 \text{ mol He}}{1.447 \text{ mol total}} = 0.86 \text{ atm}$

    $P_{Ar} = 1.00 \text{ atm} \times \dfrac{0.02503 \text{ mol Ar}}{1.447 \text{ mol total}} = 0.017 \text{ atm}$

    $P_{Ne} = 1.00 \text{ atm} \times \dfrac{0.1734 \text{ mol Ne}}{1.447 \text{ mol total}} = 0.12 \text{ atm}$

63. $P_1 = 72 \text{ cm Hg} = 720 \text{ mm Hg} \times \dfrac{1 \text{ atm}}{760 \text{ mm Hg}} = 0.95 \text{ atm}$ $P_2 = 1.00 \text{ atm}$

    $V_1 = 350 \text{ mL}$ $V_2 = ?$

    $n_1 = x$ $n_2 = 2/3 \, x$ (1/3 of gas removed)

    $T_1 = 27°C = 300 \text{ K}$ $T_2 = 600. \text{ K}$

    $PV = nRT$ or $R = \dfrac{PV}{nT} = \text{constant}$

    Thus, $\dfrac{P_1 V_1}{n_1 T_1} = \dfrac{P_2 V_2}{n_2 T_2}$

    $\dfrac{(0.95 \text{ atm})(350 \text{ mL})}{(x)(300. \text{ K})} = \dfrac{(1.00 \text{ atm})(V_2)}{(2/3 \, x)(600. \text{ K})}$

    Thus, $V_2 = 443 \text{ mL}$

Gases

64. A balloon is essentially a constant pressure container. Thus, P and n are constant.

$$PV = nRT \quad \text{or} \quad \frac{nR}{P} = \frac{V}{T} = \text{constant}$$

Thus, $\dfrac{V_1}{T_1} = \dfrac{V_2}{T_2}$

$$\frac{0.750 \text{ L}}{293 \text{ K}} = \frac{2.00 \text{ L}}{T_2}$$

$T_2 = 781 \text{ K} = 508°\text{C}$

65. $PV = nRT \quad \text{or} \quad T = \dfrac{PV}{nR}$

$P = 1.3 \times 10^9$ atm

Assume $V = 1.00$ L

$$1.00 \text{ L} \times \frac{1000 \text{ cm}^3}{1.00 \text{ L}} \times \frac{1.4 \text{ g}}{1 \text{ cm}^3} = 1400 \text{ g gas}$$

$$1400 \text{ g} \times \frac{1 \text{ mol}}{2.00 \text{ g}} = 700 \text{ mol gas}$$

Thus, $T = \dfrac{(1.3 \times 10^9 \text{ atm})(1.00 \text{ L})}{(700 \text{ mol})(0.08206 \text{ L atm mol}^{-1} \text{ K}^{-1})} = 2.26 \times 10^7 \text{ K}$

66. If we can determine the molar mass of X_4H_{10}, we can determine X.

$PV = nRT \quad \text{or} \quad n = \dfrac{PV}{RT}$

$P = 801 \text{ mm Hg} \times \dfrac{1 \text{ atm}}{760 \text{ mm Hg}} = 1.05 \text{ atm}$

$V = 30.0 \text{ cm}^3 \times \dfrac{1.00 \text{ L}}{1000 \text{ cm}^3} = 0.0300 \text{ L}$

$T = 20°\text{C} + 273 = 293 \text{ K}$

$n = \dfrac{(1.05 \text{ atm})(0.0300 \text{ L})}{(0.08206 \text{ L atm mol}^{-1} \text{ K}^{-1})(293 \text{ K})} = 1.31 \times 10^{-3} \text{ mol}$

molar mass of $X_4H_{10} = \dfrac{0.0712 \text{ g}}{1.31 \times 10^{-3} \text{ mol}} = 54.3 \text{ g/mol}$

Thus, 4(atomic mass of X) + 10(1.008) = 54.3

atomic mass of X = 11.1

This is closest to boron (B), which has an atomic mass of 10.8.

67. $PV = nRT$ or $n = \dfrac{PV}{RT}$

$P = 802 \text{ mm Hg} \times \dfrac{1 \text{ atm}}{760 \text{ mm Hg}} = 1.06 \text{ atm}$

$V = 618 \text{ cm}^3 \times \dfrac{1.00 \text{ L}}{1000 \text{ cm}^3} = 0.618 \text{ L}$

$T = 75°C + 273 = 348 \text{ K}$

$n = \dfrac{(1.06 \text{ atm})(0.618 \text{ L})}{(0.08206 \text{ L atm mol}^{-1} \text{ K}^{-1})(348 \text{ K})} = 0.0229 \text{ mol gas}$

n_{O_2} = 20% of 0.0229 = 4.58×10^{-3} mol O_2

Number of O_2 molecules = 4.58×10^{-3} mol $O_2 \times \dfrac{6.022 \times 10^{23} \text{ molecules}}{1 \text{ mol}} =$ 2.76×10^{21} O_2 molecules

68. We are looking for the subscripts for $C_xH_yO_z$

For every 100.0 g of the compound, we have

$48.6 \text{ g C} \times \dfrac{1 \text{ mol C}}{12.01 \text{ g C}} = 4.05 \text{ mol C}$

$8.18 \text{ g H} \times \dfrac{1 \text{ mol H}}{1.008 \text{ g H}} = 8.12 \text{ mol H}$

$43.2 \text{ g O} \times \dfrac{1 \text{ mol O}}{16.00 \text{ g O}} = 2.70 \text{ mol O}$

4.05 : 8.12 : 2.70 = 1.5 : 3 : 1 = 3 : 6 : 2

The empirical formula is $C_3H_6O_2$ which has a molar mass of 74.08 g/mol [3(12.01) + 6(1.008) + 2(16.00)]

Assume 1.00 L, we can solve for n

$n = \dfrac{PV}{RT} = \dfrac{(1.00 \text{ atm})(1.00 \text{ L})}{(0.08206 \text{ L atm mol}^{-1} \text{ K}^{-1})(150°C + 273)} = 0.0288 \text{ mol}$

molar mass = $\dfrac{2.13 \text{ g}}{0.0288 \text{ mol}} = 74.0$ g/mol

Thus, the molecular formula is $C_3H_6O_2$

Two possible Lewis structures include

and

Chapter 14 Liquids and Solids

1. Dipole-dipole interactions occur when molecules possessing dipole moments orient themselves so that the positive and negative ends of adjacent molecules attract each other. *Any* three polar liquids will possess dipole-dipole forces among their molecules [e.g., $CH_3Cl(l)$, $SO_2(l)$, $NO_2(l)$].

2. Dipole-dipole forces are relatively stronger at short distances; they are short-range forces. Molecules must first closely approach one another before dipole-dipole forces can cause attraction between molecules.

3. Hydrogen bonding is a special case of dipole-dipole forces that can come into play in molecules in which hydrogen atoms are directly bonded to highly electronegative atoms (such as nitrogen, oxygen, or fluorine). Such bonds are extremely polar, and because the hydrogen atom is so tiny, the dipoles of different molecules are able to approach each other much more closely. Since dipole-dipole forces are strongly distance-dependent, this makes hydrogen bonding especially strong. Three substances which would be expected to show hydrogen bonding in the liquid state are water, ammonia, and ethyl alcohol (CH_2CH_2OH).

4. Water molecules are able to form strong *hydrogen bonds* with each other. These bonds are an especially strong form of dipole-dipole forces and are only possible when hydrogen atoms are bonded to the most electronegative elements (N, O, and F). The extra strong intermolecular forces in H_2O require much higher temperatures (high energies) to be overcome in order to permit the liquid to boil. We take the fact that water has a much higher boiling point than the other hydrogen compounds of the Group 6 elements as proof that a special force is at play in water (hydrogen bonding).

5. The magnitude of a dipole-dipole interactions is very strongly dependent on the distance between the dipoles. In the solid and liquid states, the molecular dipoles are quite close together. In the vapor phase (gaseous state), however, the molecules are too far apart from one another for dipole-dipole forces to be very strong.

6. The fact that such nonpolar, monatomic atoms *can* be liquefied and solidified indicates that there must be *some* sort of intermolecular forces possible between atoms in these substances. London dispersion forces arise when a temporary (instantaneous) dipolar arrangement of charge develops as the electrons of an atom move around its nucleus. This instantaneous dipole can induce a similar dipole in a neighboring atom, leading to a momentary attraction.

7.   a. London dispersion forces (nonpolar, atoms)

      b. London dispersion forces (nonpolar molecules)

      c. dipole-dipole forces; London dispersion forces

      d. hydrogen bonding (H attached to O); London dispersion forces.

8. The boiling points increase with an increase in the molar mass of the noble gas. As the noble gas atoms increase in size, the valence electrons are farther from the nucleus. At greater distance from the nucleus, it is easy for another atom's electrons to momentarily

distort the electron cloud of the noble gas atom. As the size of the noble gas atoms increases, so does the magnitude of the London dispersion forces.

9. For a homogeneous mixture to be able to form at all, the forces between molecules of the two substances being mixed must be at least *comparable in magnitude* to the intermolecular forces within each *separate* substance. Apparently in the case of a water-ethanol mixture, the forces that exist when water and ethanol are mixed are stronger than water-water or ethanol-ethanol forces in the separate substances. This allows ethanol and water molecules to approach each other more closely in the mixture than either substance's molecules could approach a like molecule in the separate substances. There is strong hydrogen bonding in both ethanol and water.

10. Pure water is a colorless, tasteless substance that freezes to form a solid at 0°C and boils at 100°C. Water has a relatively large specific heat capacity and is able to absorb massive amounts of energy from the sun, preventing rapid or unusually large changes in temperature in the environment.

11. Water exerts its cooling effect in nature in many ways. Water, as perspiration, helps cool the human body (the evaporation of water from skin is an endothermic process; the heat required for evaporation comes from the body). Large bodies of natural water (e.g., the oceans) have a cooling effect on nearby land masses (the interior of the United States, away from the oceans, tends to be hotter than coastal regions). In industry, water is used as a coolant in *many* situations. Some nuclear power plants, for example, use water to cool the reactor core. Many office buildings are air-conditioned in summer by circulating cold water systems.

12. Ice floats because it is less dense than liquid water. Generally, the solid form of a substance is *more* dense than the liquid form. Water is an exception to this.

13. The fact that water expands when it freezes often results in broken water pipes during cold weather. The expansion of water when it freezes also makes ice float on liquid water. The expansion of a given mass of water into a larger volume upon freezing lowers the density of ice compared to liquid water. Aquatic life (and probably all life) could not exist if ice sank in water.

14. From room temperature to 100°C, adding heat to water raises the temperature of the water (as the molecules of water convert the applied heat to kinetic energy, they begin to move faster and faster). At 100°C, the boiling point, water molecules possess enough kinetic energy to escape readily from the liquid's surface (the temperature remains at 100°C until all the liquid water has been converted to steam). If heated above 100°C, steam molecules absorb additional kinetic energy and the temperature of the steam increases.

15. Sloped portions of a heating/cooling curve represent *changes in temperature* as heat is applied or removed; for example, in the cooling/heating curve shown, there are sloped portions representing the heating of ice, the heating of liquid water, and the heating of steam, as heat continues to be applied. Flat portions of such curves represent equilibrium transitions between states; for example, the flat portions in the curve shown represent the solid-liquid (melting-freezing) transition and the liquid-vapor (boiling-condensation) phase transitions.

Liquids and Solids

16. In a solid well below its melting point, the constituent particles are virtually locked in place in a regular lattice array (they can only vibrate somewhat about their mean positions). As heat energy is added to such a solid, the vibrational motions of the particles increase (the heat energy is converted to greater kinetic energies for the particles). Eventually the particles are vibrating so strongly that they are able to move apart from one another and begin to move in the more random manner of the particles in a liquid.

17. As a liquid is heated, the motions of the molecules increase as the temperature rises. As the liquid reaches its boiling point, bubbles of vapor begin to form in the liquid, which rise to the surface of the liquid and burst. As the liquid remains at its boiling point, the additional heat energy being supplied to the liquid is used to overcome attractive forces among the molecules in the liquid. As heat energy continues to be applied, more and more molecules will be moving in the right direction and with sufficient energy to escape from the liquid.

18. *Intra*molecular forces are the forces *within* a molecule itself (e.g., a covalent bond is an intramolecular force). *Inter*molecular forces are forces between or among *different* molecules. Consider liquid bromine, Br_2. *Intra*molecular forces (a covalent bond) are responsible for the fact that bromine atoms form discrete two-atom units (molecules) within the substance. *Inter*molecular forces between adjacent Br_2 molecules are responsible for the fact that the substance is a liquid at room temperature and pressure.

19. To melt a solid, or to vaporize a liquid, the molecules of the substance must be moved apart: thus, it is the *inter*molecular forces that must be overcome.

20. In ice, water molecules are in more or less regular, fixed positions in the ice crystal; strong hydrogen bonding forces exist within the ice crystal holding the water molecules together. In liquid water, enough heat has been applied that the molecules are no longer fixed in position (but are more free to roam about in the bulk of the liquid); since the water molecules are still relatively close together, strong hydrogen bonding forces still exist, however, which keep the liquid together in one place. In steam, the water molecules possess enough kinetic energy that they have escaped from the liquid; because the water molecules are very far apart in steam, and because the molecules are moving very quickly, they do not exert appreciable forces on each other (each water molecule in steam behaves independently).

21.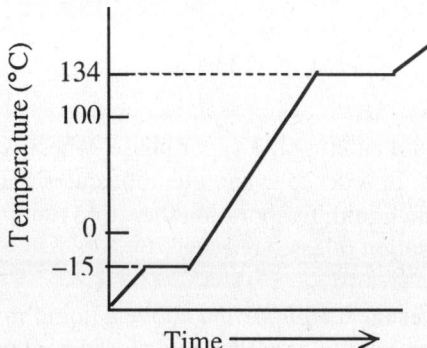

22. molar mass H_2O = 18.02

$$25.0 \text{ g } H_2O \times \frac{1 \text{ mol } H_2O}{18.02 \text{ g } H_2O} = 1.39 \text{ mol}$$

To melt the ice: $1.39 \text{ mol} \times \frac{6.02 \text{ kJ}}{1 \text{ mol}} = 8.37 \text{ kJ}$

$$37.5 \text{ g } H_2O \times \frac{1 \text{ mol } H_2O}{18.02 \text{ g } H_2O} = 2.08 \text{ mol}$$

To vaporize the liquid: $2.08 \text{ mol} \times \frac{40.6 \text{ kJ}}{1 \text{ mol}} = 84.4 \text{ kJ}$

To heat the liquid: $55.2 \text{ g} \times 4.18 \frac{\text{J}}{\text{g °C}} \times 100\text{°C} = 23{,}074 \text{ J} = 23.1 \text{ kJ}$

23. $10.0 \text{ g } H_2O = 0.555 \text{ mol } H_2O$

for steam going from 200° to 100°C:

$$Q = 10.0 \text{ g steam} \times \frac{2.03 \text{ J}}{\text{g °C}} \times 100\text{°C} = 2030 \text{ J} = 2.03 \text{ kJ}$$

released to condense the steam:

$$0.555 \text{ mol steam} \times \frac{40.6 \text{ kJ}}{1 \text{ mol}} = 22.5 \text{ kJ}$$

for liquid water going from 100° to 0°C:

$$10.0 \text{ g } H_2O \times \frac{4.184 \text{ J}}{\text{g °C}} \times 100\text{°C} = 4184 \text{ J} = 4.184 \text{ kJ}$$

to freeze the liquid water: $0.555 \text{ mol } H_2O \times \frac{6.02 \text{ kJ}}{1 \text{ mol}} = 3.34 \text{ kJ}$

for ice going from 0° to –50°C:

$$10.0 \text{ g} \times \frac{2.06 \text{ J}}{\text{g °C}} \times 50\text{°C} = 1030 \text{ J} = 1.03 \text{ kJ}$$

Total heat released = 2.03 + 22.5 + 4.184 + 3.34 + 1.03 = 33.1 kJ

24. Evaporation represents molecules of a liquid entering the vapor phase, whereas condensation is the reverse of this process. In order to evaporate, molecules must gain sufficient kinetic energy to escape from the liquid. Evaporation therefore requires an external input of energy, whereas condensation releases energy.

25. Vapor pressure is the pressure of vapor present at *equilibrium* above a liquid in a sealed container at a particular temperature. When a liquid is placed in a closed container, molecules of the liquid evaporate freely into the empty space above the liquid. As the number of molecules present in the vapor state increases with time, vapor molecules begin to rejoin the liquid state (condense). Eventually a dynamic equilibrium is reached

between evaporation and condensation in which the net number of molecules present in the vapor phase becomes *constant* with time.

26. A dynamic equilibrium is the situation in which two opposite processes are going on at the same speed, so that there is no *net* change in the system. When a liquid evaporates into the empty space above it, eventually, as more molecules accumulate in the vapor state, condensation will be occurring at the same rate as further evaporation. When evaporation and condensation are going on at the same rate, a fixed pressure of vapor will have developed, and there will be no further net change in the amount of liquid present.

27. One method uses an apparatus that consists basically of a barometer into which a volatile liquid may be injected. Since mercury is so much more dense than other liquids, the injected volatile liquid rises to the top of the mercury column in the tube and floats on top of the mercury. Since the space above the mercury is a vacuum, the volatile liquid evaporates into the empty space. As the liquid is converted to the gaseous state, the level of the mercury column drops as the pressure of vapor builds up.

28. a. H_2S — Hydrogen bonding will occur in H_2O, preventing it from evaporating as readily. In H_2S, only the relatively weak dipole-dipole forces can exist.

    b. CH_3OH — Both substances are capable of hydrogen bonding, but in H_2O there are two locations where such interactions are possible (making the hydrogen bonding in H_2O stronger than in CH_3OH).

    c. CH_3OH — Although both molecules are capable of hydrogen bonding, generally a lighter molecule is more volatile than a heavier molecule.

29. Hydrogen bonding can occur in *both* molecules. Oxygen atoms are more electronegative than nitrogen atoms, however, and the polarity of the O–H bond is considerably greater than the polarity of the N–H bond. This leads to *stronger* hydrogen bonding in liquid water than in liquid NH_3.

30. Both substances have the same molar mass. However ethyl alcohol contains a hydrogen atom directly bonded to an oxygen atom. Therefore, hydrogen bonding can exist in ethyl alcohol, whereas only weak dipole-dipole forces can exist in dimethyl ether. Dimethyl ether is more volatile; ethyl alcohol has a higher boiling point.

31. At higher altitudes the boiling temperatures of liquids are lower because there is less atmospheric pressure above the liquid. The temperature at which food cooks is determined by the temperature to which water in the food can be heated before it escapes as steam. By boiling at a lower temperature, the food will have to cook longer.

32. The boiling point is the point at which the vapor pressure of the liquid is equal to atmospheric pressure. At higher altitudes, atmospheric pressure is less, thus the boiling temperature is less.

33. A crystalline solid is a solid with a regular, repeating microscopic arrangement of its components (ions, atoms, or molecules). This highly-ordered microscopic arrangement of

the components of a crystalline solid is frequently reflected macroscopically in beautiful, regularly shaped crystals for such solids.

34. *Ionic* solids have as their fundamental particles positive and negative *ions*; a simple example is sodium chloride, in which Na^+ and Cl^- ions are held together by strong electrostatic forces.

    Molecular solids have molecules as their fundamental particles, with the molecules being held together in the crystal by dipole-dipole forces, hydrogen bonding forces, or London dispersion forces (depending on the identity of the substance); simple examples of molecular solids include ice (H_2O) and ordinary table sugar (sucrose).

    Atomic solids have simple atoms as their fundamental particles, with the atoms being held together in the crystal either by covalent bonding (as in graphite or diamond) or by metallic bonding (as in copper or other metals).

35. The fundamental particles in ionic solids are positive and negative ions. For example, the ionic solid sodium chloride consists of an alternating, regular array of Na^+ and Cl^- ions. Similarly, an ionic solid such as $CaBr_2$ consists of a regular array of Ca^{2+} ions and Br^- ions. In ionic crystals, each positive ion is surrounded by and attracted to a group of negative ions, and each negative ion is surrounded by and attracted to a group of positive ions. The fundamental particles in molecular solids are discrete molecules. Although the atoms in each molecule are held together by strong intramolecular forces (covalent bonds), the intermolecular forces in a molecular solid are not nearly as strong as in an ionic solid, which leads to molecular solids typically having relatively low melting points. Two examples of molecular solids are the common sugars glucose, $C_6H_{12}O_6$, and sucrose, $C_{12}H_{22}O_{11}$.

36. The interparticle forces in ionic solids (the ionic bond) are much stronger than the interparticle forces in molecular solids (dipole-dipole forces, London forces, etc.). The difference in intermolecular forces is most clearly shown in the great differences in melting points and boiling points between ionic and molecular solids. For example, table salt and ordinary sugar are both crystalline solids that appear very similar. Yet sugar can be melted easily in a saucepan during the making of candy, whereas even the full heat of a stove will not melt salt.

37. Ionic solids are held together by very strong electrostatic forces between the positive and negative ions. The forces are so strong that it becomes very difficult to move or displace ions from one another when an outside force is applied to the solid, and so the solid is perceived as being "hard". Molecular solids are held together by much weaker dipole-dipole forces. When an outside force seeking to deform or displace the solid is applied, these weaker forces are much easier to overcome. The overall magnitude of an electrostatic force is related (in part) to the magnitude of the charges involved. In ionic compounds the charges are "full" ionic charges and the forces are strong; in molecular solids the charges are "partial" and the forces are weaker.

38. Ionic solids consist of a crystal lattice of basically alternating positively and negatively charged ions. A given ion is surrounded by several ions of the opposite charge, all of which electrostatically attract it strongly. This pattern repeats itself throughout the crystal. The existence of these strong electrostatic forces throughout the crystal means a great deal of energy may be applied to overcome the forces and melt the solid.

Liquids and Solids

39. Strong electrostatic forces exist between oppositely charged ions in ionic solids (i.e., the attraction of a positive ion by several nearby ions of the opposite charge, and *vice versa*).

40. In solid Kr, the only forces that exist are the very weak London dispersion forces. In diamond, however, each carbon atom is held to its four nearest neighbors by strong covalent bonds.

41. An alloy represents a mixture of elements that as a whole shows metallic properties. In a substitutional alloy, some of the host metal atoms are *replaced* by other metal atoms (e.g., brass, pewter, plumber's solder). In an interstitial alloy, other small atoms occupy the spaces between the larger host metal atoms (e.g., carbon steel).

42. Alloys may be of two types: *substitutional* (in which one metal is substituted for another in the regular positions of the crystal lattice) and *interstitial* (in which a second metal's atoms fit into the empty space in a given metal's crystal lattice). The presence of atoms of a second metal in a given metal's crystal lattice changes the properties of the metal: frequently the alloy is stronger than either of the original metals because the irregularities introduced into the crystal lattice by the presence of a second metal's atoms prevent the crystal from being deformed as easily. The properties of iron may be modified by alloying with many different substances, particularly with carbon, nickel, and cobalt. Steels with relatively high carbon content are exceptionally strong, whereas steels with low carbon contents are softer, more malleable, and more ductile. Steels produced by alloying iron with nickel and cobalt are more resistant to corrosion than iron itself.

43. m

44. j

45. g

46. f

47. i

48. d

49. e

50. a

51. c

52. l

53. Dimethyl ether has the larger vapor pressure. No hydrogen bonding is possible since the O atom does not have a hydrogen atom attached. Hydrogen bonding can occur *only* when a hydrogen atom is *directly* attached to a strongly electronegative atom (such as N, O, or F). Hydrogen bonding *is* possible in 1-butanol (1-butanol contains an –OH group).

54.
   a. KBr (ionic bonding)
   b. NaCl (ionic bonding)
   c. H₂O (hydrogen bonding)

55. Evaporation of a substance is the *liquid → vapor* change of state. For every substance, a certain amount of energy is required to accomplish this change of state (heat of vaporization). Alcohol is a volatile liquid, with a relatively large heat of vaporization. Applying alcohol to a fever victim's skin causes internal heat from the body to be absorbed as the heat of vaporization of the alcohol.

56. Steel is a general term applied to alloys consisting primarily of iron, but with small amounts of other substances added. Whereas pure iron itself is relatively soft, malleable, and ductile, steels are typically much stronger and harder, and much less subject to damage.

57. Dipole-dipole interactions are typically about 1% as strong as a covalent bond. Dipole-dipole interactions represent electrostatic attractions between portions of molecules which carry only a *partial* positive or negative charge, and such forces require the molecules that are interacting to come *near* enough to each other.

58. London dispersion forces are relatively weak forces that arise among noble gas atoms and in nonpolar molecules. London forces are due to *instantaneous dipoles* that develop when one atom (or molecule) momentarily distorts the electron cloud of another atom (or molecule). London forces are typically weaker than either permanent dipole-dipole forces or covalent bonds.

59.
   a. London dispersion forces (nonpolar molecules)
   b. hydrogen bonding (H attached to N); London dispersion forces
   c. London dispersion forces (nonpolar molecules)
   d. London dispersion forces (nonpolar molecules)

60. A *volatile* liquid is one that evaporates relatively easily. Volatile liquids typically have large vapor pressures because the intermolecular forces that would tend to prevent evaporation are small.

61. In NH₃, strong hydrogen bonding can exist. In CH₄, because the molecule is nonpolar, only the relatively weak London dispersion forces exist.

Chapter 15 Solutions

1. A homogeneous mixture is a combination of two (or more) pure substances which is uniform in composition and appearance throughout. Examples of homogeneous mixtures in the real world include rubbing alcohol (70% isopropyl alcohol, 30% water) and gasoline (a mixture of hydrocarbons).

2. solvent, solute

3. When an ionic solute dissolves in water, a given ion is pulled into solution by the attractive ion-dipole force exerted by several water molecules. For example, in dissolving a positive ion, the ion is approached by the negatively charged end of several water molecules: if the attraction of the water molecules for the positive ion is stronger than the attraction of the negative ions near it in the crystal, the ion leaves the crystal and enters solution. After entering solution, the dissolved ion is surrounded completely by water molecules, which tends to prevent the ion from reentering the crystal.

4. One substance will mix with and dissolve in another substance if the intermolecular forces are similar in the two substances, so that when the mixture forms, the forces between particles in the mixture will be similar to the forces present in the separate substances. Sugar and ethyl alcohol molecules both contain polar –OH groups, which are comparable to the polar –OH structure in water. Sugar or ethyl alcohol molecules can hydrogen bond with water molecules and intermingle with them freely to form a solution. Substances like petroleum (whose molecules contain only carbon and hydrogen) are very nonpolar and cannot form interactions with polar water molecules.

5. saturated

6. unsaturated

7. variable

8. large

9. Increasing the surface area of a solid increases the amount of solid that comes in contact with the solvent. Typically, particles of a solid are broken into smaller pieces by grinding. This increases the surface to volume ratio of each particle and thus increases the amount of solid in contact with the solvent.

10. An increase in temperature means an increase in the average kinetic energy. Thus, in a warmer solution the particles of the liquid solvent are moving more rapidly. Because of this, there is an increased rate in solute-solvent interaction, which increases the rate of dissolving.

11. An increase in temperature means an increase in average kinetic energy. In a warmer solution, the solvent and solute particles are moving more rapidly. If the solute particles are gaseous, faster moving particles are more likely to have enough energy to escape from the liquid.

12.
   a. $\dfrac{5.00 \text{ g CaCl}_2}{(95.0 \text{ g H}_2\text{O} + 5.00 \text{ g CaCl}_2)} \times 100 = 5.00\% \text{ CaCl}_2$

   b. $\dfrac{1.00 \text{ g CaCl}_2}{(19.0 \text{ g H}_2\text{O} + 1.00 \text{ g CaCl}_2)} \times 100 = 5.00\% \text{ CaCl}_2$

   c. $\dfrac{15.00 \text{ g CaCl}_2}{(285 \text{ g H}_2\text{O} + 15.00 \text{ g CaCl}_2)} \times 100 = 5.00\% \text{ CaCl}_2$

   d. $\dfrac{0.00200 \text{ g CaCl}_2}{(0.0380 \text{ g H}_2\text{O} + 0.00200 \text{ g CaCl}_2)} \times 100 = 5.00\% \text{ CaCl}_2$

13. To say that a solution is $x\%$ NaCl means that 100 g of the solution would contain x g of NaCl.

   a. $11.5 \text{ g solution} \times \dfrac{6.25 \text{ g NaCl}}{100 \text{ g solution}} = 0.719 \text{ g NaCl}$

   b. $6.25 \text{ g solution} \times \dfrac{11.5 \text{ g NaCl}}{100 \text{ g solution}} = 0.719 \text{ g NaCl}$

   c. $54.3 \text{ g solution} \times \dfrac{0.91 \text{ g NaCl}}{100 \text{ g solution}} = 0.49 \text{ g NaCl}$

   d. $452 \text{ g solution} \times \dfrac{12.3 \text{ g NaCl}}{100 \text{ g solution}} = 55.6 \text{ g NaCl}$

14. $\dfrac{5.34 \text{ g KCl}}{(5.34 \text{ g KCl} + 152 \text{ g H}_2\text{O})} \times 100 = \dfrac{5.34 \text{ g}}{157.34 \text{ g}} \times 100 = 3.39\% \text{ KCl}$

15. $\dfrac{67.1 \text{ g CaCl}_2}{(67.1 \text{ g CaCl}_2 + 275 \text{ g H}_2\text{O})} \times 100 = 19.6\% \text{ CaCl}_2$

16. $285 \text{ g solution} \times \dfrac{5.00 \text{ g NaCl}}{100 \text{ g solution}} = 14.3 \text{ g NaCl}$

    $285 \text{ g solution} \times \dfrac{7.50 \text{ g Na}_2\text{CO}_3}{100.0 \text{ g solution}} = 21.4 \text{ g Na}_2\text{CO}_3$

17. g heptane = $93 \text{ g solution} \times \dfrac{5.2 \text{ g heptane}}{100. \text{ g solution}} = 4.8 \text{ g heptane}$

    g pentane = $93 \text{ g solution} \times \dfrac{2.9 \text{ g pentane}}{100. \text{ g solution}} = 2.7 \text{ g pentane}$

    g hexane = 93 g solution − 4.8 g heptane − 2.7 g pentane = 86 g hexane

18. 0.105

19. 0.221 mol Ca^{2+} ; 0.442 mol Cl^-

Solutions

20. To say that a solution has a concentration of 5 M means that in 1 L of solution (*not* solvent) there would be 5 mol of solute: to prepare such a solution one would place 5 mol of NaCl in a 1 L flask, and then add whatever amount of water is necessary so that the *total* volume would be 1 L after mixing. The NaCl will occupy some space, so the amount of water to be added will be *less* than 1.00 L.

21. Molarity = $\dfrac{\text{moles of solute}}{\text{liters of solution}}$

    a. 250 mL = 0.25 L

    $$M = \dfrac{0.50 \text{ mol KBr}}{0.25 \text{ L solution}} = 2.0 \; M$$

    b. 500 mL = 0.500 L

    $$M = \dfrac{0.50 \text{ mol KBr}}{0.500 \text{ L solution}} = 1.0 \; M$$

    c. 750 mL = 0.75 L

    $$M = \dfrac{0.50 \text{ mol KBr}}{0.75 \text{ L solution}} = 0.67 \; M$$

    d. $M = \dfrac{0.50 \text{ mol KBr}}{1.0 \text{ L solution}} = 0.50 \; M$

22. Molarity = $\dfrac{\text{moles of solute}}{\text{liters of solution}}$

    a. Molar mass of $CuCl_2$ = 134.45 g   125 mL = 0.125 L

    $$4.25 \text{ g } CuCl_2 \times \dfrac{1 \text{ mol}}{134.45 \text{ g}} = 0.0316 \text{ mol } CuCl_2$$

    $$M = \dfrac{0.0316 \text{ mol } CuCl_2}{0.125 \text{ L solution}} = 0.253 \; M$$

    b. Molar mass of $NaHCO_3$ = 84.01 g   11.3 mL = 0.0113 L

    $$0.101 \text{ g } NaHCO_3 \times \dfrac{1 \text{ mol}}{84.01 \text{ g}} = 0.00120 \text{ mol } NaHCO_3$$

    $$M = \dfrac{0.00120 \text{ mol } NaHCO_3}{0.0113 \text{ L solution}} = 0.106 \; M$$

    c. Molar mass of Na_2CO_3 = 105.99 g

    $$52.9 \text{ g } Na_2CO_3 \times \dfrac{1 \text{ mol}}{105.99 \text{ g}} = 0.499 \text{ mol } Na_2CO_3$$

    $$M = \dfrac{0.499 \text{ mol } Na_2CO_3}{1.15 \text{ L solution}} = 0.434 \; M$$

d.   Molar mass of KOH = 56.11 g   1.5 mL = 0.0015 L

$$0.14 \text{ mg KOH} \times \frac{1 \text{ g}}{10^3 \text{ mg}} \times \frac{1 \text{ mol}}{56.11 \text{ g}} = 2.50 \times 10^{-6} \text{ mol KOH}$$

$$M = \frac{2.50 \times 10^{-6} \text{ mol KOH}}{0.0015 \text{ L solution}} = 1.67 \times 10^{-3} \, M = 1.7 \times 10^{-3} \, M$$

23.   molar mass of KNO_3 = 101.1 g   225 mL = 0.225 L

$$45.3 \text{ g KNO}_3 \times \frac{1 \text{ mol}}{101.1 \text{ g}} = 0.448 \text{ mol KNO}_3$$

$$M = \frac{0.448 \text{ mol KNO}_3}{0.225 \text{ L solution}} = 1.99 \, M$$

24.   molar mass of I_2 = 253.8 g   225 mL = 0.225 L

$$5.15 \text{ g I}_2 \times \frac{1 \text{ mol}}{253.8 \text{ g}} = 0.0203 \text{ mol I}_2$$

$$M = \frac{0.0203 \text{ mol I}_2}{0.225 \text{ L solution}} = 0.0902 \, M$$

25.   molar mass of $FeCl_3$ = 162.2 g

$$1.01 \text{ g FeCl}_3 \times \frac{1 \text{ mol FeCl}_3}{162.2 \text{ g FeCl}_3} = 0.00623 \text{ mol FeCl}_3$$

10.0 mL = 0.0100 L

$$M = \frac{0.00623 \text{ mol FeCl}_3}{0.0100 \text{ L solution}} = 0.623 \, M$$

Since one mole of $FeCl_3$ contains one mole of Fe^{3+} and three moles of Cl^-, the solution is 0.623 M in Fe^{3+} and 3(0.623) = 1.87 M in Cl^-

26.   molar mass of NaOH = 40.00 g

$$495 \text{ g NaOH} = \frac{1 \text{ mol}}{40.00 \text{ g}} = 12.4 \text{ mol NaOH}$$

$$M = \frac{12.4 \text{ mol NaOH}}{20.0 \text{ L solution}} = 0.619 \, M$$

27.   a.   molar mass of HNO_3 = 63.02 g   127 mL = 0.127 L

$$0.127 \text{ L solution} \times \frac{0.105 \text{ mol HNO}_3}{1.00 \text{ L solution}} = 0.0133 \text{ mol HNO}_3$$

$$0.0133 \text{ mol HNO}_3 \times \frac{63.02 \text{ g HNO}_3}{1 \text{ mol HNO}_3} = 0.838 \text{ g HNO}_3$$

Solutions

155

b. molar mass of NH_3 = 17.03 g   155 mL = 0.155 L

$$0.155 \text{ L solution} \times \frac{15.1 \text{ mol NH}_3}{1.00 \text{ L solution}} = 2.34 \text{ mol NH}_3$$

$$2.34 \text{ mol NH}_3 \times \frac{17.03 \text{ g NH}_3}{1 \text{ mol NH}_3} = 39.9 \text{ g NH}_3$$

c. molar mass KSCN = 97.19 g

$$2.51 \text{ L solution} \times \frac{2.01 \times 10^{-3} \text{ mol KSCN}}{1.00 \text{ L solution}} = 5.05 \times 10^{-3} \text{ mol KSCN}$$

$$5.05 \times 10^{-3} \text{ mol KSCN} \times \frac{97.19 \text{ g KSCN}}{1 \text{ mol KSCN}} = 0.491 \text{ g KSCN}$$

d. molar mass of HCl = 36.46 g   12.2 mL = 0.0122 L

$$0.0122 \text{ L solution} \times \frac{2.45 \text{ mol HCl}}{1.00 \text{ L solution}} = 0.0299 \text{ mol HCl}$$

$$0.0299 \text{ mol HCl} \times \frac{36.46 \text{ g HCl}}{1 \text{ mol HCl}} = 1.09 \text{ g HCl}$$

28. molar mass of NH_4Cl = 53.49 g   450 mL = 0.450 L

$$0.450 \text{ L solution} \times \frac{0.251 \text{ mol NH}_4\text{Cl}}{1 \text{ L solution}} = 0.113 \text{ mol NH}_4\text{Cl}$$

$$0.113 \text{ mol NH}_4\text{Cl} \times \frac{53.49 \text{ g NH}_4\text{Cl}}{1 \text{ mol NH}_4\text{Cl}} = 6.04 \text{ g NH}_4\text{Cl}$$

29. a. 10.2 mL = 0.0102 L

$$0.0102 \text{ L} \times \frac{0.451 \text{ mol AlCl}_3}{1.00 \text{ L}} \times \frac{1 \text{ mol Al}^{3+}}{1 \text{ mol AlCl}_3} = 4.60 \times 10^{-3} \text{ mol Al}^{3+}$$

$$0.0102 \text{ L} \times \frac{0.451 \text{ mol AlCl}_3}{1.00 \text{ L}} \times \frac{3 \text{ mol Cl}^-}{1 \text{ mol AlCl}_3} = 1.38 \times 10^{-2} \text{ mol Cl}^-$$

b. $$5.51 \text{ L} \times \frac{0.103 \text{ mol Na}_3\text{PO}_4}{1.00 \text{ L}} \times \frac{3 \text{ mol Na}^+}{1 \text{ mol Na}_3\text{PO}_4} = 1.70 \text{ mol Na}^+$$

$$5.51 \text{ L} \times \frac{0.103 \text{ mol Na}_3\text{PO}_4}{1.00 \text{ L}} \times \frac{1 \text{ mol PO}_4^{3-}}{1 \text{ mol Na}_3\text{PO}_4} = 0.568 \text{ mol PO}_4^{3-}$$

c. 1.75 mL = 0.00175 L

$$0.00175 \text{ L} \times \frac{1.25 \text{ mol CuCl}_2}{1.00 \text{ L}} \times \frac{1 \text{ mol Cu}^{2+}}{1 \text{ mol CuCl}_2} = 2.19 \times 10^{-3} \text{ mol Cu}^{2+}$$

$$0.00175 \text{ L} \times \frac{1.25 \text{ mol CuCl}_2}{1.00 \text{ L}} \times \frac{2 \text{ mol Cl}^-}{1 \text{ mol CuCl}_2} = 4.38 \times 10^{-3} \text{ mol Cl}^-$$

d. 25.2 mL = 0.0252

$$0.0252 \text{ L} \times \frac{0.00157 \text{ mol Ca(OH)}_2}{1.00 \text{ L}} \times \frac{1 \text{ mol Ca}^{2+}}{1 \text{ mol Ca(OH)}_2} = 3.96 \times 10^{-5} \text{ mol Ca}^{2+}$$

$$0.0252 \text{ L} \times \frac{0.00157 \text{ mol Ca(OH)}_2}{1.00 \text{ L}} \times \frac{2 \text{ mol OH}^-}{1 \text{ mol Ca(OH)}_2} = 7.91 \times 10^{-5} \text{ mol OH}^-$$

30. 250. mL = 0.250 L

$$0.250 \text{ L solution} \times \frac{0.100 \text{ mol AgNO}_3}{1.00 \text{ L solution}} = 0.0250 \text{ mol AgNO}_3$$

molar mass $AgNO_3$ = 169.9 g

$$0.0250 \text{ mol AgNO}_3 \times \frac{169.9 \text{ g AgNO}_3}{1 \text{ mol AgNO}_3} = 4.25 \text{ g AgNO}_3$$

31. $M_1 \times V_1 = M_2 \times V_2$

    a. $M_1 = 0.251\ M$ $M_2 = ?$

         $V_1 = 125$ mL $V_2 = 250. + 125 = 375$ mL

$$M_2 = \frac{(0.251\ M)(125 \text{ mL})}{(375 \text{ mL})} = 0.0837\ M$$

    b. $M_1 = 0.499\ M$ $M_2 = ?$

         $V_1 = 445$ mL $V_2 = 445 + 250. = 695$ mL

$$M_2 = \frac{(0.499\ M)(445 \text{ mL})}{(695 \text{ mL})} = 0.320\ M$$

    c. $M_1 = 0.101\ M$ $M_2 = ?$

         $V_1 = 5.25$ L $V_2 = 5.25 + 0.250 = 5.50$ L

$$M_2 = \frac{(0.101\ M)(5.25 \text{ L})}{(5.50 \text{ L})} = 0.0964\ M$$

    d. $M_1 = 14.5\ M$ $M_2 = ?$

         $V_1 = 11.2$ mL $V_2 = 11.2 + 250. = 261.2$ mL

$$M_2 = \frac{(14.5\ M)(11.2 \text{ mL})}{(261.2 \text{ mL})} = 0.622\ M$$

32. $M_1 \times V_1 = M_2 \times V_2$

    HCl: $M_1 = 3.0\ M$ $M_2 = 12.1\ M$

                $V_1 = 225$ mL $V_2 = ?$

$$V_2 = \frac{(3.0\ M)(225 \text{ mL})}{(12.1\ M)} = 55.8 \text{ mL} = 56 \text{ mL}$$

Solutions 157

HNO_3: $\quad M_1 = 3.0\ M \quad\quad\quad M_2 = 15.9\ M$
$\quad\quad\quad V_1 = 225\ mL \quad\quad V_2 = ?$

$$V_2 = \frac{(3.0\ M)(225\ mL)}{(15.9\ M)} = 42.45\ mL = 42\ mL$$

H_2SO_4: $\quad M_1 = 3.0\ M \quad\quad\quad M_2 = 18.0\ M$
$\quad\quad\quad V_1 = 225\ mL \quad\quad V_2 = ?$

$$V_2 = \frac{(3.0\ M)(225\ mL)}{(18.0\ M)} = 37.5\ mL = 38\ mL$$

$HC_2H_3O_2$: $\quad M_1 = 3.0\ M \quad\quad\quad M_2 = 17.5\ M$
$\quad\quad\quad V_1 = 225\ mL \quad\quad V_2 = ?$

$$V_2 = \frac{(3.0\ M)(225\ mL)}{(17.5\ M)} = 38.6\ mL = 39\ mL$$

H_3PO_4: $\quad M_1 = 3.0\ M \quad\quad\quad M_2 = 14.9\ M$
$\quad\quad\quad V_1 = 225\ mL \quad\quad V_2 = ?$

$$V_2 = \frac{(3.0\ M)(225\ mL)}{(14.9\ M)} = 45.3\ mL = 45\ mL$$

33. $M_1 \times V_1 = M_2 \times V_2$

$M_1 = 3.02\ M \quad\quad\quad\quad M_2 = 0.150\ M$
$V_1 = ? \quad\quad\quad\quad\quad\quad V_2 = 125\ mL = 0.125\ L$

$$V_1 = \frac{(0.150\ M)(0.125\ L)}{(3.02\ M)} = 0.00621\ L = 6.21\ mL$$

The student could prepare her solution by transferring 6.21 mL of the 3.02 M NaOH solution from a pipet or buret to a 125-mL volumetric flask, and then adding distilled water to the calibration mark of the flask.

34. $M_1 \times V_1 = M_2 \times V_2$

$M_1 = 0.200\ M \quad\quad\quad\quad M_2 = 0.150\ M$
$V_1 = 500.\ mL = 0.500\ L \quad\quad V_2 = ?$

$$V_2 = \frac{(0.200\ M)(0.500\ L)}{(0.150\ M)} = 0.667\ L = 667\ mL$$

Therefore: 667 − 500. = 167 mL of water must be added.

35. 27.2 mL = 0.0272 L $\quad\quad\quad$ 25.0 mL = 0.0250 L

$$\text{mol } AgNO_3 = 0.0272\ L\ \text{solution} \times \frac{0.104\ \text{mol } AgNO_3}{1.00\ L\ \text{solution}} = 0.002829\ \text{mol } AgNO_3$$

$$0.002829 \text{ mol AgNO}_3 \times \frac{1 \text{ mol Cl}^-}{1 \text{ mol AgNO}_3} = 0.002829 \text{ mol Cl}^-$$

$$M = \frac{0.002829 \text{ mol Cl}^-}{0.0250 \text{ L}} = 0.113 \; M$$

36. $Ba(NO_3)_2(aq) + Na_2SO_4(aq) \rightarrow BaSO_4(s) + 2NaNO_3(aq)$

    12.5 mL = 0.0125 L

    $$\text{moles Ba(NO}_3)_2 = 0.0125 \text{ L} \times \frac{0.15 \text{ mol Ba(NO}_3)_2}{1.00 \text{ L}} = 1.88 \times 10^{-3} \text{ mol Ba(NO}_3)_2$$

    From the balanced chemical equation for the reaction, if 1.88×10^{-3} mol $Ba(NO_3)_2$ are to be precipitated, then 1.88×10^{-3} mol Na_2SO_4 will be needed.

    $$1.88 \times 10^{-3} \text{ mol Na}_2SO_4 \times \frac{1.00 \text{ L}}{0.25 \text{ mol Na}_2SO_4} = 0.0075 \text{ L required} = 7.5 \text{ mL}$$

37. 36.2 mL = 0.0362 L   37.5 mL = 0.0375 L

    Since each formula unit of $CaCO_3$ contains one Ca^{2+} ion, and since each Na_2CO_3 formula unit contains one CO_3^{2-} ion, we can say that

    $$\text{mol Ca}^{2+} = 0.0362 \text{ L CaCl}_2 \times \frac{0.158 \text{ mol CaCl}_2}{1 \text{ L CaCl}_2} = 0.00572 \text{ mol Ca}^{2+}$$

    $$\text{mol CO}_3^{2-} = 0.0375 \text{ L Na}_2CO_3 \times \frac{0.149 \text{ mol Na}_2CO_3}{1 \text{ L Na}_2CO_3} = 0.00559 \text{ mol CO}_3^{2-}$$

    Since one Ca^{2+} reacts with one CO_3^{2-}, Na_2CO_3 is the limiting reactant. Since 0.00559 mol CO_3^{2-} reacts, 0.00559 mol of $CaCO_3$ will form.

    molar mass $CaCO_3$ = 100.1 g

    $$0.00559 \text{ mol CaCO}_3 \times \frac{100.1 \text{ g CaCO}_3}{1 \text{ mol CaCO}_3} = 0.560 \text{ g CaCO}_3$$

38. $Pb(NO_3)_2(aq) + K_2CrO_4(aq) \rightarrow PbCrO_4(s) + 2KNO_3(aq)$

    molar masses: $Pb(NO_3)_2$, 331.2 g; $PbCrO_4$, 323.2 g

    $$1.00 \text{ g Pb(NO}_3)_2 \times \frac{1 \text{ mol Pb(NO}_3)_2}{331.2 \text{ g Pb(NO}_3)_2} = 0.003019 \text{ mol Pb(NO}_3)_2$$

    25.0 mL = 0.0250 L

    $$0.0250 \text{ L solution} \times \frac{1.00 \text{ mol K}_2CrO_4}{1.00 \text{ L solution}} = 0.0250 \text{ mol K}_2CrO_4$$

    $Pb(NO_3)_2$ is the limiting reactant: 0.003019 mol $PbCrO_4$ will form.

    $$0.003019 \text{ mol PbCrO}_4 \times \frac{323.2 \text{ g PbCrO}_4}{1 \text{ mol PbCrO}_4} = 0.976 \text{ g PbCrO}_4$$

Solutions

39. $HCl(aq) + NaOH(aq) \rightarrow NaCl(aq) + H_2O(l)$

    25.0 mL = 0.0250 L

    $0.0250 \text{ L} \times \dfrac{0.150 \text{ mol NaOH}}{1.00 \text{ L}} = 0.00375 \text{ mol NaOH}$

    $0.00375 \text{ mol NaOH} \times \dfrac{1 \text{ mol HCl}}{1 \text{ mol NaOH}} = 0.00375 \text{ mol HCl}$

    $0.00375 \text{ mol HCl} \times \dfrac{1 \text{ L solution}}{0.200 \text{ mol HCl}} = 0.01875 \text{ L} = 18.8 \text{ mL}$

40. $HCl(aq) + NaOH(aq) \rightarrow NaCl(aq) + H_2O(l)$

    50.0 mL = 0.0500 L   48.7 mL = 0.0487 L

    $0.0500 \text{ L} \times \dfrac{0.104 \text{ mol HCl}}{1.00 \text{ L solution}} = 0.00520 \text{ mol HCl}$

    $0.00520 \text{ mol HCl} \times \dfrac{1 \text{ mol NaOH}}{1 \text{ mol HCl}} = 0.00520 \text{ mol NaOH}$

    $M = \dfrac{0.00520 \text{ mol NaOH}}{0.0487 \text{ L}} = 0.107 \, M$

41. a. $NaOH(aq) + HC_2H_3O_2(aq) \rightarrow NaC_2H_3O_2(aq) + H_2O(l)$

    25.0 mL = 0.0250 L

    $0.0250 \text{ L} \times \dfrac{0.154 \text{ mol } HC_2H_3O_2}{1.00 \text{ L}} = 0.00385 \text{ mol } HC_2H_3O_2$

    $0.00385 \text{ mol } HC_2H_3O_2 \times \dfrac{1 \text{ mol NaOH}}{1 \text{ mol } HC_2H_3O_2} = 0.00385 \text{ mol NaOH}$

    $0.00385 \text{ mol NaOH} \times \dfrac{1.00 \text{ L}}{1.00 \text{ mol NaOH}} = 0.00385 \text{ L} = 3.85 \text{ mL NaOH}$

    b. $HF(aq) + NaOH(aq) \rightarrow NaF(aq) + H_2O(l)$

    35.0 mL = 0.0350 L

    $0.0350 \text{ L} \times \dfrac{0.102 \text{ mol HF}}{1.00 \text{ L}} = 0.00357 \text{ mol HF}$

    $0.00357 \text{ mol HF} \times \dfrac{1 \text{ mol HF}}{1 \text{ mol NaOH}} = 0.00357 \text{ mol NaOH}$

    $0.00357 \text{ mol NaOH} \times \dfrac{1.00 \text{ L}}{1.00 \text{ mol NaOH}} = 0.00357 \text{ L} = 3.57 \text{ mL}$

    c. $H_3PO_4(aq) + 3NaOH(aq) \rightarrow Na_3PO_4(aq) + 3H_2O(l)$

    10.0 mL = 0.0100 L

$$0.0100 \text{ L} \times \frac{0.143 \text{ mol } H_3PO_4}{1.00 \text{ L}} = 0.00143 \text{ mol } H_3PO_4$$

$$0.00143 \text{ mol } H_3PO_4 \times \frac{3 \text{ mol NaOH}}{1 \text{ mol } H_3PO_4} = 0.00429 \text{ mol NaOH}$$

$$0.00429 \text{ mol NaOH} \times \frac{1.00 \text{ L}}{1.00 \text{ mol NaOH}} = 0.00429 \text{ L} = 4.29 \text{ mL}$$

d. $H_2SO_4(aq) + 2NaOH(aq) \rightarrow Na_2SO_4(aq) + 2H_2O(l)$

35.0 mL = 0.0350 L

$$0.0350 \text{ L} \times \frac{0.220 \text{ mol } H_2SO_4}{1.00 \text{ L}} = 0.00770 \text{ mol } H_2SO_4$$

$$0.00770 \text{ mol } H_2SO_4 \times \frac{2 \text{ mol NaOH}}{1 \text{ mol } H_2SO_4} = 0.0154 \text{ mol NaOH}$$

$$0.0154 \text{ mol NaOH} \times \frac{1.00 \text{ L}}{1.00 \text{ mol NaOH}} = 0.0154 \text{ L} = 15.4 \text{ mL}$$

42. When H_2SO_4 reacts with OH^-, the reaction is

$H_2SO_4(aq) + 2OH^-(aq) \rightarrow 2H_2O(l) + SO_4^{2-}(aq)$

Since each mol of H_2SO_4 provides *two* moles of H^+ ion, it is only necessary to take *half* a mole of H_2SO_4 to provide *one* mole of H^+ ion. The equivalent weight of H_2SO_4 is thus half the molar mass.

43. 1.53 equivalents OH^- ion are needed to react with 1.53 equivalents of H^+ ion. By *definition*, one equivalent of OH^- ion exactly neutralizes one equivalent of H^+ ion.

44. $N = \dfrac{\text{number of equivalents of solute}}{\text{number of liters of solution}}$

   a. equivalent weight NaOH = molar mass NaOH = 40.00 g

   $$0.113 \text{ g NaOH} \times \frac{1 \text{ equiv NaOH}}{40.00 \text{ g}} = 2.83 \times 10^{-3} \text{ equiv NaOH}$$

   10.2 mL = 0.0102 L

   $$N = \frac{2.83 \times 10^{-3} \text{ equiv}}{0.0102 \text{ L}} = 0.277 \, N$$

   b. equivalent weight $Ca(OH)_2 = \dfrac{\text{molar mass}}{2} = \dfrac{74.10 \text{ g}}{2} = 37.05 \text{ g}$

   $$12.5 \text{ mg} \times \frac{1 \text{ g}}{10^3 \text{ mg}} \times \frac{1 \text{ equiv}}{37.05 \text{ g}} = 3.37 \times 10^{-4} \text{ equiv } Ca(OH)_2$$

   100. mL = 0.100 L

Solutions

$$N = \frac{3.37 \times 10^{-3} \text{ equiv}}{0.100 \text{ L}} = 3.37 \times 10^{-3} \, N$$

c. equivalent weight H_2SO_4 = $\frac{\text{molar mass}}{2}$ = $\frac{98.09 \text{ g}}{2}$ = 49.05 g

$$12.4 \text{ g} \times \frac{1 \text{ equiv}}{49.05 \text{ g}} = 0.253 \text{ equiv } H_2SO_4$$

155 mL = 0.155 L

$$N = \frac{0.253 \text{ equiv}}{0.155 \text{ L}} = 1.63 \, N$$

45. a. $0.134 \, M$ NaOH × $\frac{1 \text{ equiv NaOH}}{1 \text{ mol NaOH}}$ = $0.134 \, N$ NaOH

b. $0.00521 \, M$ $Ca(OH)_2$ × $\frac{2 \text{ equiv } Ca(OH)_2}{1 \text{ mol } Ca(OH)_2}$ = $0.0104 \, N$ $Ca(OH)_2$

c. $4.42 \, M$ H_3PO_4 × $\frac{3 \text{ equiv } H_3PO_4}{1 \text{ mol } H_3PO_4}$ = $13.3 \, N$ H_3PO_4

46. molar mass H_3PO_4 = 98.0 g

$$35.2 \text{ g } H_3PO_4 \times \frac{1 \text{ mol } H_3PO_4}{98.0 \text{ g } H_3PO_4} = 0.3592 \text{ mol } H_3PO_4$$

$$M = \frac{0.3592 \, H_3PO_4}{1.00 \text{ L}} = 0.3592 \, M = 0.359 \, M$$

$$0.3592 \, M \, H_3PO_4 \times \frac{3 \text{ equiv } H_3PO_4}{1 \text{ mol } H_3PO_4} = 1.08 \, N$$

47. $H_2SO_4(aq) + 2NaOH(aq) \rightarrow Na_2SO_4(aq) + 2H_2O(l)$

$0.145 \, M$ NaOH – $0.145 \, N$ NaOH   56.2 mL = 0.0562 L

$$0.0562 \text{ L NaOH} \times \frac{0.145 \text{ equiv}}{1.00 \text{ L}} = 0.00815 \text{ equiv NaOH}$$

0.00815 equiv NaOH requires 0.00815 equiv H_2SO_4 to react.

$$0.00815 \text{ equiv } H_2SO_4 \times \frac{1.00 \text{ L}}{0.172 \text{ equiv}} = 0.0474 \text{ L} = 47.4 \text{ mL } H_2SO_4 \text{ solution}$$

48. $2NaOH(aq) + H_2SO_4(aq) \rightarrow Na_2SO_4(aq) + 2H_2O(l)$

For the $0.125 \, N$ H_2SO_4:

$N_{acid} \times V_{acid} = N_{base} \times V_{base}$

$(0.125 \, N) \times (24.2 \text{ mL}) = (0.151 \, N) \times (V_{base})$

V_{base} = 20.0 mL of the $0.151 \, N$ NaOH solution needed

For the 0.125 M H_2SO_4:

Since each H_2SO_4 formula unit produces two H^+ ions, the normality of this solution will be twice its molarity

0.125 M H_2SO_4 = 0.250 N H_2SO_4

N_{acid} x V_{acid} = N_{base} x V_{base}

(0.250 N) x (24.1 mL) = (0.151 N) x (V_{base})

V_{base} = 39.9 mL of the 0.151 N NaOH solution needed

49. Colligative properties are properties of a solution that depend only on the number, not the identity, of the solute particles.

50. For a solution to boil, bubbles must form in the solution. Solute particles block water from entering these bubbles. It is not the nature of these particles that matters, but the number of the particles; thus, it is a colligative property.

51. Antifreeze solution is a concentrated aqueous solution that has a lower freezing point than water. It will also have a higher boiling point than water (a solute in water both lowers the freezing point and raises the boiling point).

52. millimol $CoCl_2$ = 50.0 mL x 0.250 M $CoCl_2$ = 12.5 millimol $CoCl_2$

This contains 12.5 millimol Co^{2+} and 25.0 millimol Cl^-

millimol $NiCl_2$ = 25.0 mL x 0.350 M $NiCl_2$ = 8.75 millimol $NiCl_2$

This contains 8.75 millimol Ni^{2+} and 17.5 millimol Cl^-

Total millimol Cl^- after mixing = 25.0 + 17.5 = 42.5 millimol Cl^-

Total volume after mixing = 50.0 mL + 25.0 mL = 75.0 mL

$$M_{cobalt(II)\ ion} = \frac{12.5\ \text{millimol}\ Co^{2+}}{75.0\ \text{mL}} = 0.167\ M$$

$$M_{nickel(II)\ ion} = \frac{8.75\ \text{millimol}\ Ni^{2+}}{75.0\ \text{mL}} = 0.117\ M$$

$$M_{chloride\ ion} = \frac{42.5\ \text{millimol}\ Cl^-}{75.0\ \text{mL}} = 0.567\ M$$

53. $AgNO_3(s) + NaCl(aq) \rightarrow AgCl(s) + NaNO_3(aq)$

molar masses: $AgNO_3$, 169.9 g; $AgCl$, 143.4 g

$$10.0\ g\ AgNO_3 \times \frac{1\ mol\ AgNO_3}{169.9\ g\ AgNO_3} = 0.05886\ mol\ AgNO_3$$

50. mL = 0.050 L

$$0.050\ L \times \frac{1.0 \times 10^{-2}\ mol\ NaCl}{1.00\ L} = 0.00050\ mol\ NaCl$$

Solutions

NaCl is the limiting reactant. 0.00050 mol AgCl form.

$$0.00050 \text{ mol AgCl} \times \frac{143.4 \text{ g AgNO}_3}{1 \text{ mol}} = 0.072 \text{ g AgCl (72 mg)}$$

Since 1 mol AgNO$_3$ contains 1 mol Ag$^+$, the mol Ag$^+$ remaining in solution =

0.05886 − 0.00050 = 0.05836 mol AgNO$_3$

0.05836 mol AgNO$_3$ = 0.05836 mol Ag$^+$

$$M_{\text{silver ion}} = \frac{0.05836 \text{ mol Ag}^+}{0.050 \text{ L}} = 1.167 \, M = 1.2 \, M$$

54. Ba(NO$_3$)$_2$(aq) + H$_2$SO$_4$(aq) → BaSO$_4$(s) + 2HNO$_3$(aq)

   37.5 mL = 0.0375 L

   $$0.0375 \text{ L} \times \frac{0.221 \text{ mol H}_2\text{SO}_4}{1.00 \text{ L}} = 0.00829 \text{ mol H}_2\text{SO}_4$$

   Since the coefficients of Ba(NO$_3$)$_2$ and H$_2$SO$_4$ in the balanced chemical equation for the reaction are both *one*, then 0.00829 mol of Ba^{2+} ion will be precipitated from the solution as BaSO$_4$.

   molar mass BaSO$_4$ = 233.4 g

   $$0.00829 \text{ mol BaSO}_4 \times \frac{233.4 \text{ g BaSO}_4}{1 \text{ mol BaSO}_4} = 1.93 \text{ g BaSO}_4 \text{ precipitate}$$

55. molar mass H$_2$O = 18.0 g

   1.0 L water = 1.0 x 10^3 mL water ≅ 1.0 x 10^3 g water

   $$1.0 \times 10^3 \text{ g H}_2\text{O} \times \frac{1 \text{ mol H}_2\text{O}}{18.0 \text{ g H}_2\text{O}} = 56 \text{ mol H}_2\text{O}$$

56. molar mass CaCl$_2$ = 111.0 g

   $$14.2 \text{ g CaCl}_2 \times \frac{1 \text{ mol CaCl}_2}{111.0 \text{ g CaCl}_2} = 0.128 \text{ mol CaCl}_2$$

   50.0 mL = 0.0500 L

   $$M = \frac{0.128 \text{ mol CaCl}_2}{0.0500 \text{ L}} = 2.56 \, M$$

57. $M_1 \times V_1 = M_2 \times V_2$

   a. $M_1 = 0.200 \, M$ $M_2 = ?$

         $V_1 = 125$ mL $V_2 = 125 + 150. = 275$ mL

   $$M_2 = \frac{(0.200 \, M)(125 \text{ mL})}{(275 \text{ mL})} = 0.0909 \, M$$

b. $M_1 = 0.250\ M$ $M_2 = ?$

$V_1 = 155$ mL $V_2 = 155 + 150. = 305$ mL

$$M_2 = \frac{(0.250\ M)(155\ \text{mL})}{(305\ \text{mL})} = 0.127 M$$

c. $M_1 = 0.250\ M$ $M_2 = ?$

$V_1 = 0.500$ L $= 500.$ mL $V_2 = 500. + 150. = 650.$ mL

$$M_2 = \frac{(0.250\ M)(500.\ \text{mL})}{(650.\ \text{mL})} = 0.192\ M$$

d. $M_1 = 18.0\ M$ $M_2 = ?$

$V_1 = 15$ mL $V_2 = 15 + 150. = 165$ mL

$$M_2 = \frac{(18.0\ M)(15\ \text{mL})}{(165\ \text{mL})} = 1.6\ M$$

58. a. $0.50\ M\ HC_2H_3O_2 \times \dfrac{1\ \text{equiv}\ HC_2H_3O_2}{1\ \text{mol}\ HC_2H_3O_2} = 0.50\ N\ HC_2H_3O_2$

b. $0.00250\ M\ H_2SO_4 \times \dfrac{2\ \text{equiv}\ H_2SO_4}{1\ \text{mol}\ H_2SO_4} = 0.00500\ N\ H_2SO_4$

c. $0.10\ M\ KOH \times \dfrac{1\ \text{equiv}\ KOH}{1\ \text{mol}\ KOH} = 0.10\ N\ KOH$

59. $N_{acid} \times V_{acid} = N_{base} \times V_{base}$

$N_{acid} \times (10.0\ \text{mL}) = (3.5 \times 10^{-2}\ N)(27.5\ \text{mL})$

$N_{acid} = 9.6 \times 10^{-2}\ N\ HNO_3$

Chapter 16 Acids and Bases

1. Acids were recognized primarily from their sour taste. Bases were recognized from their bitter taste and slippery feel on skin.

2. In the Arrhenius definition, an acid is a substance which produces hydrogen ions (H^+) when dissolved in water, whereas a base is a substance which produces hydroxide ions (OH^-) in aqueous solution. These definitions proved to be too restrictive since the only base permitted was hydroxide ion, and the only solvent permitted was water.

3. A Brønsted-Lowry acid is a molecule or ion capable of providing a proton to some other species; acids are *proton donors*. A Brønsted-Lowry base is a molecule or ion capable of receiving a proton from some other species; bases are *proton acceptors*. It is the *transfer of protons* that characterizes the Brønsted-Lowry model for acids and bases.

4. The members of a conjugate acid-base pair differ from each other by one proton (one hydrogen ion, H^+). For example, CH_3COOH (acetic acid), differs from its conjugate base, CH_3COO^- (acetate ion), by a single H^+ ion.

$$CH_3COOH(aq) \rightarrow CH_3COO^-(aq) + H^+(aq)$$

5. A Brønsted-Lowry acid converts into its conjugate base in water (aqueous) solution by transferring a proton to a water molecule (forming an H_3O^+ ion). The portion of the original acid molecule or ion that *remains* after the proton leaves is the conjugate base of the original acid. In this process, water behaves as a Brønsted-Lowry base, since it receives a proton from the acid.

6. When an acid is dissolved in water, the hydronium ion (H_3O^+) is formed. The hydronium ion is the conjugate *acid* of water (H_2O).

7. a. H_2SO_4 and SO_4^{2-} do *not* represent a conjugate acid-base pair, since they differ from each other by more than one proton. The conjugate base of H_2SO_4 is the HSO_4^- ion; the conjugate acid of SO_4^{2-} is also the HSO_4^- ion.

    b. $H_2PO_4^-$ and HPO_4^{2-} represent a conjugate acid-base pair.

    c. $HClO_4$ and Cl^- are *not* a conjugate acid-base pair since they differ in the number of oxygen atoms present. The perchlorate ion, ClO_4^- is the conjugate base of the acid $HClO_4$; the conjugate acid of Cl^- is HCl.

    d. NH_4^+ and NH_2^- are *not* a conjugate acid-base pair since they differ from each other by more than one proton. NH_3 is the conjugate base of NH_4^+ and also the conjugate acid of NH_2^-.

8. a. NH_3 (base), NH_4^+ (acid); H_2O (acid), OH^- (base)

    b. PO_4^{3-} (base), HPO_4^{2-} (acid); H_2O (acid), OH^- (base)

    c. $C_2H_3O_2^-$ (base), $HC_2H_3O_2$ (acid); H_2O (acid), OH^- (base)

9. The conjugate *acid* of the species indicated would have *one additional proton*:

    a. H_2SO_4

    b. HSO_3^-

c. $HClO_4$

d. H_3PO_4

10. The conjugate *bases* of the species indicated would have *one less proton*:

    a. CO_3^{2-}

    b. HPO_4^{2-}

    c. Cl^-

    d. SO_4^{2-}

11. a. $NH_3^+ + H_2O \rightleftharpoons NH_4^+ + OH^-$

    b. $NH_2^- + H_2O \rightarrow NH_3 + OH^-$

    c. $O^{2-} + H_2O \rightarrow OH^- + OH^-$

    d. $F^- + H_2O \rightleftharpoons HF + OH^-$

12. a. $HClO_4 + H_2O \rightarrow ClO_4^- + H_3O^+$

    b. $HC_2H_3O_2 + H_2O \rightleftharpoons C_2H_3O_2^- + H_3O^+$

    c. $HSO_3^- + H_2O \rightleftharpoons SO_3^{2-} + OH^-$

    d. $HBr + H_2O \rightarrow Br^- + H_3O^+$

13. A strong acid is one for which the equilibrium in water lies far to the right. A strong acid is almost completely converted to its conjugate base when dissolved in water. A strong acid's anion (its conjugate base) must be very poor at attracting, or holding onto, protons. A regular arrow ($\rightarrow$) rather than a double arrow ($\rightleftharpoons$) is used when writing an equation for the dissociation of a strong acid to indicate this.

14. To say that an acid is *weak* in aqueous solution means that the acid does not easily transfer protons to water (and does not fully ionize). If an acid does not lose protons easily, then the acid's anion must be a strong attractor of protons (good at holding on to protons).

15. A strong acid is one which loses its protons easily and fully ionizes in water; this means that the acid's conjugate base must be poor at attracting and holding on to protons, and is therefore a relatively weak base. A weak acid is one which resists loss of its protons and does not ionize well in water; this means that the acid's conjugate base attracts and holds onto protons tightly and is a relatively strong base.

16. H_2SO_4 (sulfuric): $H_2SO_4 + H_2O \rightarrow HSO_4^- + H_3O^+$

    HCl (hydrochloric): $HCl + H_2O \rightarrow Cl^- + H_3O^+$

    HNO_3 (nitric): $HNO_3 + H_2O \rightarrow NO_3^- + H_3O^+$

    $HClO_4$ (perchloric): $HClO_4 + H_2O \rightarrow ClO_4^- + H_3O^+$

17. Acids that are *strong* have relatively weak conjugate bases.

    a. CH_3COO^- is a relatively strong base; CH_3COOH is a weak acid.

Acids and Bases

    b. F^- is a relatively strong base; HF is a weak acid.

    c. HS^- is a relatively strong base; H_2S is a weak acid.

    d. Cl^- is a very weak base; HCl is a strong acid.

18. Bases that are *weak* have relatively strong conjugate acids:

    a. SO_4^{2-} is a moderately weak base; HSO_4^- is a moderately strong acid

    b. Br^- is a very weak base; HBr is a strong acid

    c. CN^- is a fairly strong base; HCN is a weak acid

    d. CH_3COO^- is a fairly strong base; CH_3COOH is a weak acid

19. For example, HCO_3^- can behave as an acid if it reacts with something that more strongly gains protons than does HCO_3^- itself. For example, HCO_3^- would behave as an acid when reacting with hydroxide ion (a much stronger base).

$$HCO_3^-(aq) + OH^-(aq) \rightarrow CO_3^{2-}(aq) + H_2O(l)$$

On the other hand, HCO_3^- would behave as a base when reacted with something that more readily loses protons than does HCO_3^- itself. For example, HCO_3^- would behave as a base when reacting with hydrochloric acid (a much stronger acid).

$$HCO_3^-(aq) + HCl(aq) \rightarrow H_2CO_3(aq) + Cl^-(aq)$$

For $H_2PO_4^-$, similar equations can be written:

$$H_2PO_4^-(aq) + OH^-(aq) \rightarrow HPO_4^{2-}(aq) + H_2O(l)$$

$$H_2PO_4^-(aq) + H_3O^+(aq) \rightarrow H_3PO_4(aq) + H_2O(l)$$

20. $H_2O + H_2O \rightleftharpoons H_3O^+ + OH^-$ $K_w = [H_3O^+][OH^-] = 1.0 \times 10^{-14}$

        $H_2O \rightleftharpoons H^+ + OH^-$ $K_w = [H^+][OH^-] = 1.0 \times 10^{-14}$

21. The hydrogen ion concentration and the hydroxide ion concentration of water are *not* independent of each other: they are related by the equilibrium

$$H_2O(l) \rightleftharpoons H^+(aq) + OH^-(aq)$$

for which $K_w = [H^+][OH^-] = 1.0 \times 10^{-14}$ at 25°C.

If the concentration of one of these ions is increased by addition of a reagent producing H^+ or OH^-, then the concentration of the complementary ion will have to decrease so that the value of K_w will hold true. So if an acid is added to a solution, the concentration of hydroxide ion in the solution will decrease to a lower value. Similarly, if a base is added to a solution, then the concentration of hydrogen ion will have to decrease to a lower value.

22. $K_w = [H^+][OH^-] = 1.0 \times 10^{-14}$ at 25°C

$$[H^+] = \frac{1.0 \times 10^{-14}}{[OH^-]}$$

a. $[H^+] = \dfrac{1.0 \times 10^{-14}}{3.99 \times 10^{-5} M} = 2.5 \times 10^{-10}$ M; solution is basic

b. $[H^+] = \dfrac{1.0 \times 10^{-14}}{2.91 \times 10^{-9} M} = 3.4 \times 10^{-6}$ M; solution is acidic

c. $[H^+] = \dfrac{1.0 \times 10^{-14}}{7.23 \times 10^{-2} M} = 1.4 \times 10^{-13}$ M; solution is basic

d. $[H^+] = \dfrac{1.0 \times 10^{-14}}{9.11 \times 10^{-7} M} = 1.1 \times 10^{-8}$ M; solution is basic

23. $[OH^-] = \dfrac{1.0 \times 10^{-14}}{[H^+]}$

a. $[OH^-] = \dfrac{1.0 \times 10^{-14}}{1.00 \times 10^{-7} M} = 1.0 \times 10^{-7}$ M; solution is neutral

b. $[OH^-] = \dfrac{1.0 \times 10^{-14}}{7.00 \times 10^{-7} M} = 1.4 \times 10^{-8}$ M; solution is acidic

c. $[OH^-] = \dfrac{1.0 \times 10^{-14}}{7.00 \times 10^{-1} M} = 1.4 \times 10^{-14}$ M; solution is acidic

d. $[OH^-] = \dfrac{1.0 \times 10^{-14}}{5.99 \times 10^{-6} M} = 1.7 \times 10^{-9}$ M; solution is acidic

24. a. $[H^+] = 1.2 \times 10^{-3}$ M is more acidic

b. $[H^+] = 2.6 \times 10^{-6}$ M is more acidic

c. $[H^+] = 0.000010$ M is more acidic

25. a. $[H^+] = 1.04 \times 10^{-8}$ M is more basic

b. $[OH^-] = 4.49 \times 10^{-6}$ M is more basic

c. $[OH^-] = 6.01 \times 10^{-7}$ M is more basic

26. Because the concentrations of $[H^+]$ and $[OH^-]$ in aqueous solutions tend to be expressed in scientific notation, and since these numbers have negative exponents for their powers of ten, it tends to be clumsy to make comparisons between different concentrations of these ions (see questions 24 and 25 above). The pH scale converts such numbers into "ordinary" numbers between 0 and 14 which can be more easily compared. The pH of a solution is defined as the *negative* of the base 10 logarithm of the hydrogen ion concentration, pH = $-\log[H^+]$.

27. household ammonia (pH 12); blood (pH 7-8); milk (pH 6-7); vinegar (pH 3); lemon juice (pH 2-3); stomach acid (pH 2)

28. Since 2.33×10^{-6} has three significant figures, the pH should be expressed to the third decimal place. The figure *before* the decimal place in a pH is *not* one of the significant

Acids and Bases

digits: the figure before the decimal place is related to the *power of ten* (exponent) of the concentration.

29. The pH of a solution is defined as the *negative* of the logarithm of the hydrogen ion concentration, $pH = -\log[H^+]$. Mathematically, the *negative sign* in the definition causes the pH to *decrease* as the hydrogen ion concentration *increases*.

30. $pH = -\log[H^+]$

    a. $pH = -\log[0.00100\ M] = 3.000$; solution is acidic

    b. $pH = -\log[2.19 \times 10^{-4}\ M] = 3.660$; solution is acidic

    c. $pH = -\log[9.18 \times 10^{-11}\ M] = 10.037$; solution is basic

    d. $pH = -\log[4.71 \times 10^{-7}\ M] = 6.327$; solution is acidic

31. $pOH = -\log[OH^-]$ $pH = 14.00 - pOH$

    a. $pOH = -\log[1.00 \times 10^{-7}\ M] = 7.000$

    $pH = 14.00 - 7.000 = 7.00$; solution is neutral

    b. $pOH = -\log[4.59 \times 10^{-13}\ M] = 12.338$

    $pH = 14.00 - 12.338 = 1.66$; solution is acidic

    c. $pOH = -\log[1.04 \times 10^{-4}\ M] = 3.983$

    $pH = 14.00 - 3.983 = 10.02$; solution is basic

    d. $pOH = -\log[7.00 \times 10^{-1}\ M] = 0.155$

    $pH = 14.00 - 0.155 = 13.85$; solution is basic

32. $pH = 14 - pOH$

    a. $pH = 14.00 - 4.32 = 9.68$; solution is basic

    b. $pH = 14.00 - 8.90 = 5.10$; solution is acidic

    c. $pH = 14.00 - 1.81 = 12.19$; solution is basic

    d. $pH = 14.00 - 13.1 = 0.9$; solution is acidic

33. a. $[H^+] = 1.00 \times 10^{-7}\ M$

    $[OH^-] = \dfrac{1.0 \times 10^{-14}}{1.00 \times 10^{-7}\ M} = 1.0 \times 10^{-7}\ M$

    $pH = -\log[1.0 \times 10^{-7}\ M] = 7.00$

    $pOH = 14.00 - 7.00 = 7.00$

    b. $[OH^-] = 4.39 \times 10^{-5}\ M$

    $[H^+] = \dfrac{1.0 \times 10^{-14}}{4.39 \times 10^{-5}\ M} = 2.28 \times 10^{-10}\ M = 2.3 \times 10^{-10}\ M$

$$\text{pH} = -\log[2.28 \times 10^{-10}\ M] = 9.64$$
$$\text{pOH} = 14.00 - 9.64 = 4.36$$

    c. $[H^+] = 4.29 \times 10^{-11}\ M$

$$[OH^-] = \frac{1.0 \times 10^{-14}}{4.29 \times 10^{-11}\ M} = 2.33 \times 10^{-4}\ M = 2.3 \times 10^{-4}\ M$$
$$\text{pH} = -\log[4.29 \times 10^{-11}\ M] = 10.368$$
$$\text{pOH} = 14.00 - 10.368 = 3.63$$

    d. $[OH^-] = 7.36 \times 10^{-2}\ M$

$$[H^+] = \frac{1.0 \times 10^{-14}}{7.36 \times 10^{-2}\ M} = 1.36 \times 10^{-13}\ M = 1.4 \times 10^{-13}\ M$$
$$\text{pH} = -\log[1.36 \times 10^{-13}\ M] = 12.87$$
$$\text{pOH} = 14.00 - 12.87 = 1.13$$

34. $[H^+] = \{inv\}\{log\}[-pH]$

    a. $[H^+] = \{inv\}\{log\}[-1.04] = 9.2 \times 10^{-2}\ M$

    b. $[H^+] = \{inv\}\{log\}[-13.1] = 8 \times 10^{-14}\ M$

    c. $[H^+] = \{inv\}\{log\}[-5.99] = 1.0 \times 10^{-6}\ M$

    d. $[H^+] = \{inv\}\{log\}[-8.62] = 2.4 \times 10^{-9}\ M$

35.   a. $\text{pH} = 14.00 - 3.91 = 10.09$

           $[H^+] = \{inv\}\{log\}[-10.09] = 8.1 \times 10^{-11}\ M$

    b. $\text{pH} = 14.00 - 12.56 = 1.44$

           $[H^+] = \{inv\}\{log\}[-1.44] = 3.6 \times 10^{-2}\ M$

    c. $\text{pH} = 14.00 - 1.15 = 12.85$

           $[H^+] = \{inv\}\{log\}[-12.85] = 1.4 \times 10^{-13}\ M$

    d. $\text{pH} = 14.00 - 8.77 = 5.23$

           $[H^+] = \{inv\}\{log\}[-5.23] = 5.9 \times 10^{-6}\ M$

36.   a. $\text{pOH} = 14.00 - 5.12 = 8.88$

           $[H^+] = \{inv\}\{log\}[-5.12] = 7.6 \times 10^{-6}\ M$

           $[OH^-] = \{inv\}\{log\}[-8.88] = 1.3 \times 10^{-9}\ M$

    b. $\text{pH} = 14.00 - 5.12 = 8.88$

           $[H^+] = \{inv\}\{log\}[-8.88] = 1.3 \times 10^{-9}\ M$

           $[OH^-] = \{inv\}\{log\}[-5.12] = 7.6 \times 10^{-6}\ M$

Acids and Bases

    c.   pOH = 14.00 − 7.00 = 7.00

          $[H^+] = [OH^-]$ {inv}{log}[−7.00] = $1.0 \times 10^{-7}\ M$

    d.   pH = 14.00 − 13.00 = 1.00

          $[H^+] =$ {inv}{log}[−1.00] = $1.0 \times 10^{-1}\ M$

          $[OH^-] =$ {inv}{log}[−13.00] = $1.0 \times 10^{-13}\ M$

37.   We can measure the pH of a solution with a pH meter, pH paper, and indicators. The pH meter is the most accurate.

38.   Indicators are substances that turn different colors when H^+ is present than when H^+ is absent. Consider the equation: $HIn \rightleftharpoons H^+ + In^-$. The substance HIn is a different color than the substance In^-.

39.   Effectively *no* molecules of HCl remain in solution. HCl is a strong acid for which the equilibrium with water lies far to the right. All the HCl molecules originally dissolved in the water will ionize.

40.   The solution contains water molecules, H_3O^+ ions (protons), and NO_3^- ions. Because HNO_3 is a strong acid, which is completely ionized in water, there are no HNO_3 molecules present.

41.   a.   HCl is a strong acid and completely ionized so

          $[H^+] = 1.04 \times 10^{-4}\ M$

          pH = $-\log[1.04 \times 10^{-4}] = 3.983$

    b. HNO_3 is a strong acid and completely ionized so

          $[H^+] = 0.00301\ M$

          pH = $-\log[0.00301] = 2.521$

    c. $HClO_4$ is a strong acid and completely ionized so

          $[H^+] = 5.41 \times 10^{-4}\ M$

          pH = $-\log[5.41 \times 10^{-4}] = 3.267$

    d. HNO_3 is a strong acid and completely ionized so

          $[H^+] = 6.42 \times 10^{-2}\ M$

          pH = $-\log[6.42 \times 10^{-2}] = 1.192$

42.   a.   HCl is a strong acid and completely ionized so

          $[H^+] = 0.00010\ M$ and pH = 4.00

    b. HNO_3 is a strong acid and completely ionized so

          $[H^+] = 0.0050\ M$ and pH = 2.30

c. HClO₄ is a strong acid and completely ionized so

$[H^+] = 4.21 \times 10^{-5}$ M and pH = 4.376

d. HNO₃ is a strong acid and completely ionized so

$[H^+] = 6.33 \times 10^{-3}$ M and pH = 2.199

43. HCl + NaOH → NaNO₃ + H₂O

Thus, we have a 1:1 mole ratio between HCl and NaOH

$$100.0 \text{ mL HCl} \times \frac{1.00 \text{ L}}{1000 \text{ mL}} \times \frac{0.50 \text{ mol HCl}}{1.00 \text{ L}} \times \frac{1 \text{ mol NaOH}}{1 \text{ mol HCl}} \times \frac{1.00 \text{ L}}{0.10 \text{ mol NaOH}} \times \frac{1000 \text{ mL}}{1.0 \text{ L}} = 5.0 \times 10^2 \text{ mL NaOH}$$

44. HNO₃ + NaOH → NaNO₃ + H₂O

Thus, we have a 1:1 mol ratio between HNO₃ and NaOH

$$26.5 \text{ mL NaOH} \times \frac{1.00 \text{ L}}{1000 \text{ mL}} \times \frac{0.20 \text{ mol NaOH}}{1 \text{ L}} \times \frac{1 \text{ mol HNO}_3}{1 \text{ mol NaOH}} = 0.0053 \text{ mol HNO}_3 \text{ required}$$

$$50.0 \text{ mL} \times \frac{1.00 \text{ L}}{1000 \text{ mL}} = 0.0500 \text{ L}$$

$$[HNO_3] = \frac{0.0053 \text{ mol HNO}_3}{0.0500 \text{ L}} = 0.11 \, M \text{ HNO}_3$$

45. A buffered solution is one that resists a change in its pH even when a strong acid or base is added to it. A solution is buffered by the presence of the combination of a weak acid and its conjugate base.

46. A buffered solution consists of a mixture of a weak acid and its conjugate base; one example of a buffered solution is a mixture of acetic acid (CH₃COOH) and sodium acetate (NaCH₃COO).

47. The conjugate *base* component of the buffer mixture is capable of combining with any strong acid that might be added to the buffered solution. For the example of acetate ion (C₂H₃O₂⁻) given in the solution to question 46, the equation is

HCl(*aq*) + C₂H₃O₂⁻(*aq*) → HC₂H₃O₂(*aq*) + Cl⁻(*aq*)

48. The weak *acid* component of a buffered solution is capable of reacting with added strong base. For example, using the buffered solution given as an example in Question 46, acetic acid would consume added sodium hydroxide as follows:

CH₃COOH(*aq*) + NaOH(*aq*) → NaCH₃COO(*aq*) + H₂O(*l*)

Acetic acid *neutralizes* the added NaOH and prevents it from having much effect on the overall pH of the solution.

49. a. not a buffer: although HCl and Cl⁻ are conjugates, HCl is *not* a weak acid.

b. a buffer: CH₃COOH and CH₃COO⁻ are conjugates

Acids and Bases

        c.   a buffer: H_2S and HS^- are conjugates

        d.   not a buffer: S^{2-} (of Na_2S) is not the conjugate base of H_2S

50.   added NaOH: $CH_3COOH + OH^- \rightarrow CH_3COO^- + H_2O$

        added HCl: $CH_3COO^- + H_3O^+ \rightarrow CH_3COOH + H_2O$

        added NaOH: $H_2S + OH^- \rightarrow HS^- + H_2O$

        added HCl: $HS^- + H_3O^+ \rightarrow H_2S + H_2O$

51.   In whatever solvent is used, a Brønsted-Lowry acid will still be a proton donor. In liquid ammonia, HCl would still be an acid through the following equation

$$HCl + NH_3 \rightarrow Cl^- + NH_4^+$$

in which the proton is transferred from HCl to NH_3. Similarly, OH^- would still be a base (proton acceptor) in liquid ammonia as indicated in the equation

$$OH^- + NH_3 \rightarrow H_2O + NH_2^-$$

in which OH^- could receive a proton from ammonia.

52.   *a*, *c*, and *d* represent acidic solutions; *b* represents a *basic* solution because there is more hydroxide ion present than there is hydrogen ion.

53.   *a*, *b*, and *d* represent basic solutions; *c* represents an *acidic* solution because there is less hydroxide ion than hydrogen ion.

54.   *a*, *c*, and *e* represent strong acids; *b* and *d* are typical *weak* acids.

55.   a. H_2O and OH^- represent a conjugate acid-base pair (H_2O is the acid, having one more proton than the base, OH^-).

        b. H_2SO_4 and SO_4^{2-} are *not* a conjugate acid-base pair (they differ by *two* protons). The conjugate base of H_2SO_4 is HSO_4^-; the conjugate acid of SO_4^{2-} is also HSO_4^-.

        c. H_3PO_4 and $H_2PO_4^-$ represent a conjugate acid-base pair (H_3PO_4 is the acid, having one more proton than the base $H_2PO_4^-$).

        d. $HC_2H_3O_2$ and $C_2H_3O_2^-$ represent a conjugate acid-base pair ($HC_2H_3O_2$ is the acid, having one more proton than the base $C_2H_3O_2^-$)

56.   a. CH_3NH_2 (base), $CH_3NH_3^+$ (acid); H_2O (acid), OH^- (base)

        b. CH_3COOH (acid), CH_3COO^- (base); NH_3 (base), NH_4^+ (acid)

        c.   HF (acid), F^- (base); NH_3 (base), NH_4^+ (acid)

57.   Bases that are *weak* have relatively strong conjugate acids:

        a. F^- is a relatively strong base; HF is a weak acid.

        b. Cl^- is a very weak base; HCl is a strong acid.

c. HSO_4^- is a very weak base; H_2SO_4 is a strong acid.

d. NO_3^- is a very weak base; HNO_3 is a strong acid.

58. a. $[OH^-] = 0.0000032\ M$ is more basic

b. $[OH^-] = 1.54 \times 10^{-8}\ M$ is more basic

c. $[OH^-] = 4.02 \times 10^{-7}\ M$ is more basic

59. mol NaOH = mol OH^-

mol HNO_3 = mol H^+

$$\text{mol } OH^- = 90.0\ mL \times \frac{1.00\ L}{1000\ mL} \times \frac{0.400\ mol}{1.00\ L} = 0.0360\ mol\ OH^-$$

$$\text{mol } H^+ = 30.0\ mL \times \frac{1.00\ L}{1000\ mL} \times \frac{1.75\ mol}{1.00\ L} = 0.0525\ mol\ H^+$$

The net ionic reaction is $H^+ + OH^- \rightarrow H_2O$
Thus, H^+ is in excess,
and the leftover H^+ = 0.0525 mol – 0.0360 mol = 0.0165 mol H^+

Total volume = 90.0 mL + 30.0 mL + 240.0 mL = 360.0 mL = 0.360 L

$$[H^+] = \frac{0.0165\ mol}{0.360\ L} = 0.0458\ M,\ \text{acidic}$$

$pH = -\log[H^+] = 1.339$

$pH + pOH = 14.00$, $pOH = 12.66$

$[OH^-] = 10^{-pOH} = 2.18 \times 10^{-13}\ M$

60. $1\ \text{drop} \times \dfrac{1.00\ mL}{20\ \text{drops}} = 0.0500\ mL$ (per drop)

$$\text{mol HCl} = \text{mol } H^+ = 0.0500\ mL \times \frac{1.00\ L}{1000\ mL} \times \frac{0.200\ mol}{1.00\ L} = 1.00 \times 10^{-5}\ mol\ H^+$$

total volume = 1.05 mL = 0.00105 L

$$[H^+] = \frac{1.00 \times 10^{-5}\ mol}{0.00105\ L} = 9.52 \times 10^{-3}\ M$$

$[H^+]$ in pure water (25°C) = $1.00 \times 10^{-7}\ M$

Thus, the increase is $\dfrac{9.52 \times 10^{-3}}{1.00 \times 10^{-7}} = 95{,}200$ times as great.

That is, there is 95,200 times the concentration of H^+ in the solution when 1 drop of 0.200 M HCl was added (compared to pure water).

61. $pH = -\log[H^+] = -\log(1.00) = 0.00$

Doubling the volume, we have the same number of moles of H^+ that were originally present in 100.0 mL of solution. That is,

Acids and Bases

$$\text{moles H}^+ = 100.0 \text{ mL} \times \frac{1.00 \text{ L}}{1000 \text{ mL}} \times \frac{1.00 \text{ mol}}{1.00 \text{ L}} = 0.100 \text{ mol H}^+ \text{ originally present.}$$

But the volume of the solution is now 200.0 mL rather than 100.0 mL. Thus, [H$^+$] changes, as does pH.

$$\text{pH} = -\log\left(\frac{0.100 \text{ mol}}{0.200 \text{ L}}\right) = 0.301$$

Similarly, doubling the volume again, we get

$$\text{pH} = -\log\left(\frac{0.100 \text{ mol}}{0.400 \text{ L}}\right) = 0.602$$

The maximum value for the pH is 7.00 because pure water contributes 1.00×10^{-7} M H$^+$.

Chapter 17 Equilibrium

1. The hydrogen-hydrogen bond of H_2 and the bromine-bromine bond of Br_2 must break. Two hydrogen-bromine bonds (one in each of the HBr molecules) must form.

2. The nitrogen-nitrogen triple bond in N_2 and the three hydrogen-hydrogen bonds (in the three H_2 molecules) must be broken. Six nitrogen-hydrogen bonds must form (in the two ammonia molecules).

3. The collision model pictures chemical reactions as taking place only when the reactant molecules *physically collide* with one another, with enough energy to break bonds in the reactant molecules. Not all collisions possess enough energy to break bonds in the reactant molecules. A minimum energy, the *activation energy* (E_a) is needed for a collision to result in reaction. If molecules do not possess this minimum energy when they collide, they just bounce off one another without reacting.

4. The activation energy is the minimum energy two colliding molecules must possess in order for the collision to result in reaction. If molecules do not possess energies equal to or greater than E_a, a collision between these molecules will not result in a reaction.

5. A catalyst is a substance which speeds up a reaction without being consumed in the reaction (the full amount of catalyst used is still present after the reaction is complete). Catalysts work by providing an alternative pathway by which a reaction can take place: this alternative pathway has a lower activation energy.

6. Living cells contain biological catalysts called *enzymes*. Such enzymes are necessary to speed up the complicated biochemical processes that must occur in cells. Such processes would be too slow to sustain life at room temperature if such catalysts were not present.

7. In a heterogeneous reaction, reactants are in two phases. Reactions involving only one phase are homogeneous reactions.

8. For a chemical reaction to occur, the reactants must collide with one another with an energy greater than or equal to the activation energy. As the temperature is increased, the average kinetic energy is increased which results in the particles moving faster. Therefore, collisions will occur more frequently and a larger percentage of these collisions will have an energy greater than or equal to the activation energy. Thus, the number of collisions resulting in a reaction increases with temperature, that is, the rate of reaction is increased.

9. Grain dust becomes airborne when grain is poured into a silo. This dust consists of tiny particles, which exposes a large amount of surface area to the oxygen in the air.

10. In an equilibrium system, two opposing processes are going on at the same time and at the same speed. There is no net change in a system at equilibrium with time. Every time one process occurs, the opposite process occurs at the same time elsewhere in the system. A simple equilibrium might exist for the populations of two similar size towns connected by highway. Assuming there is no great attraction in one town compared to the other, we might assume the populations of the two towns would remain constant with time as

Equilibrium

individual people drive between them, but in such a way that new people are arriving in the first town from the second town as residents of the first town leave for the second town.

11. A state of equilibrium is attained when two opposing processes are exactly balanced. The development of a vapor pressure above a liquid in a closed container is an example of a physical equilibrium. Any chemical reaction which appears to "stop" before completion serves as an example of a chemical equilibrium.

12. Chemical reactions are reversible if they can occur in either direction (as written from left to right, or the reverse of this). In principle, all chemical reactions are microscopically reversible. In practice, many (though by no means all) reactions are favored greatly in one direction over the other.

13. Chemical equilibrium occurs when two *opposing* chemical reactions reach the *same speed* in a closed system. When a state of chemical equilibrium has been reached, the concentrations of reactants and products present in the system remain *constant* with time, and the reaction appears to "stop." A chemical reaction that reaches a state of equilibrium is indicated by using a double arrow ($\rightleftharpoons$). The points of the double arrow point in opposite directions, indicating that two opposite processes are going on.

14. Once a system has reached equilibrium the net concentration of product no longer increases because molecules of product already present react to form the original reactants. Although there is no *net* change in the number of product molecules present at any one time, this is not to say that the *same* product molecules are always present.

15. Although we recognize a state of chemical equilibrium by the fact that the concentrations of reactants and products no longer change with time, the lack of change results from the fact that two *opposing* processes are going on at the same time with the same rate (not because the reaction has truly "stopped"). Further reaction in the forward direction is canceled out by an equal extent of reaction in the reverse direction. The reaction is still proceeding, but the opposite reaction is also proceeding at the same rate.

16. The equilibrium constant represents a ratio of the concentration of products present at the point of equilibrium to the concentration of reactants present, with the concentration of each species raised to the power of its coefficient in the balanced chemical equation for the reaction. For a general reaction

    $$a\text{A} + b\text{B} \rightleftharpoons c\text{C} + d\text{D}$$

    the equilibrium constant, K, has the algebraic form

    $$K = \frac{[\text{C}]^c [\text{D}]^d}{[\text{A}]^a [\text{B}]^b}$$

    where square brackets indicate the concentrations of the substances in moles per liter (molarity, M).

17. The equilibrium constant is a *ratio* of the concentration of products to the concentration of reactants, with all concentrations measured at equilibrium. Depending on how much reactant a particular experiment was begun with, there may be different absolute amounts

of reactants and products present at equilibrium, but the *ratio* will always be the same for a given reaction at a given temperature. For example, the ratios (4/2) and (6/3) are different absolutely in terms of the numbers involved, but each of these ratios has the *value* of 2.

18. a. $K = \dfrac{[NCl_3]^2}{[N_2][Cl_2]^3}$

    b. $K = \dfrac{[HI]^2}{[H_2][I_2]}$

    c. $K = \dfrac{[N_2H_4]}{[N_2][H_2]^2}$

19. a. $K = \dfrac{[CH_3OH]}{[CO][H_2]^2}$

    b. $K = \dfrac{[NO]^2[O_2]}{[NO_2]^2}$

    c. $K = \dfrac{[PBr_3]^4}{[P_4][Br_2]^6}$

20. $CH_3OH(g) \rightleftharpoons CH_2O(g) + H_2(g)$

    $K = \dfrac{[CH_2O][H_2]}{[CH_3OH]} = \dfrac{[0.441\ M][0.0331\ M]}{[0.00215\ M]} = 6.79$

21. $N_2(g) + 3H_2(g) \rightleftharpoons 2NH_3(g)$

    $K = \dfrac{[NH_3]^2}{[N_2][H_2]^3} = \dfrac{[0.34\ M]^2}{[4.9 \times 10^{-4}\ M][2.1 \times 10^{-3}\ M]^3} = 2.5 \times 10^{10}$

22. $N_2(g) + O_2(g) \rightleftharpoons 2NO(g)$

    $K = \dfrac{[NO]^2}{[N_2][O_2]} = \dfrac{[4.7 \times 10^{-4}\ M]^2}{[0.041\ M][0.0078\ M]} = 6.9 \times 10^{-4}$

23. A *homogeneous* equilibrium system is a system in which all the substances present are in the same physical state. An example is

    $N_2(g) + O_2(g) \rightleftharpoons 2NO(g)$

Equilibrium

Heterogeneous equilibrium systems are those involving substances in more than one physical state (e.g., mixtures of liquids and gases, gases and solids, etc.). Examples are:

$$BaCO_3(s) \rightleftharpoons BaO(s) + CO_2(g)$$

$$NH_3(g) + HCl(g) \rightleftharpoons NH_4Cl(s)$$

24. Equilibrium constants represent ratios of the *concentrations* of products and reactants present at the point of equilibrium. The *concentration* of a pure solid or of a pure liquid is constant and is determined by the density of the solid or liquid. For example, suppose you had a liter of water. Within that liter of water are 55.5 mol of water (the number of moles of water that is contained in one liter of water *does not vary*).

25.
   a. $K = [H_2O(g)][CO_2(g)]$
   b. $K = [CO_2]$
   c. $K = \dfrac{1}{[O_2]^3}$

26.
   a. $K = [Br_2]^3[N_2]$
   b. $K = \dfrac{[H_2O]}{[H_2]}$
   c. $K = [CO_2]$

27. Le Châtelier's principle states that when a change is imposed on a system at equilibrium, the position of the equilibrium shifts in a direction that tends to reduce the effect of the change.

28. When an additional amount of one of the reactants is added to an equilibrium system, the system shifts to the right and adjusts so as to use up some of the added reactant. This results in a net *increase* in the amount of product, compared to the equilibrium system before the additional reactant was added. The numerical *value* of the equilibrium constant does *not* change when a reactant is added: the concentrations of all reactants and products adjust until the correct value of K is once again achieved.

29. When the volume of an equilibrium system involving gaseous substances is decreased suddenly, the pressure in the container increases. Reaction will occur, shifting in the direction that gives the smaller number of gas molecules, to reduce this increase in pressure.

30. If heat is applied to an endothermic reaction (i.e., the temperature is raised), the equilibrium is shifted to the right. More product will be present at equilibrium than if the temperature had not been increased. The value of K increases.

31.
   a. shifts right (system reacts to get rid of excess O_2)
   b. shifts right (system reacts to replace water)
   c. shifts left (system reacts to replace gaseous ammonia)

32. a. no change (water is in the *liquid* state)
    b. (assuming the system is warm enough to convert the dry ice to the gaseous state) the equilibrium will shift to the left.
    c. shifts to right
    d. shifts to right

33. Since the reaction is endothermic, heat is effectively a reactant. Performing the reaction at a lower temperature (removing heat from the system) would shift the reaction to the left.

34. For an *endo*thermic reaction, an increase in temperature will shift the position of equilibrium to the right (toward products).

35. the production of dextrose will be favored

36. A large equilibrium constant means that the concentration of products is large, compared to the concentration of remaining reactants. The position of the equilibrium lies far to the right. Reactions with numerically large equilibrium constants are greatly favored as a source of product. When we calculated the theoretical yield for a reaction in earlier chapters, we tacitly assumed that the reaction had a large equilibrium constant.

37. A small equilibrium constant means that the concentration of products is small, compared to the concentration of reactants. The position of equilibrium lies far to the left. Reactions with very small equilibrium constants are generally not very useful as a source of the products, unless Le Châtelier's principle can be applied to shift the position of equilibrium to the point where a sufficient amount of product can be isolated.

38. $K = \dfrac{[NCl_3]^2}{[N_2][Cl_2]^3} = \dfrac{[1.9 \times 10^{-1}\ M]^2}{[1.4 \times 10^{-3}\ M][4.3 \times 10^{-4}\ M]^3} = 3.2 \times 10^{11}$

39. $K = \dfrac{[O_3]^2}{[O_2]^3}$

    $1.8 \times 10^{-7} = \dfrac{[O_3]^2}{[0.0012\ M]^3}$

    $[O_3]^2 = 1.8 \times 10^{-7} \times [0.0012]^3 = 3.11 \times 10^{-16}$

    $[O_3] = 1.8 \times 10^{-8}\ M$

40. $K = [CO_2] = [2.1 \times 10^{-3}\ M] = 2.1 \times 10^{-3}$

41. $K = \dfrac{[NO]^2}{[N_2][O_2]}$

Equilibrium

$$1.71 \times 10^{-3} = \frac{[NO]^2}{[0.0342\ M][0.0342\ M]}$$

$[NO]^2 = 1.71 \times 10^{-3} \times [0.0342] \times [0.0342] = 2.00 \times 10^{-6}$

$[NO] = 1.41 \times 10^{-3}\ M$

42. $K = \dfrac{[NO_2]^2}{[N_2O_4]} = 8.1 \times 10^{-3}$

$$8.1 \times 10^{-3} = \frac{[0.0021\ M]^2}{[N_2O_4]}$$

$[N_2O_4] = 5.4 \times 10^{-4}\ M$

43. When a crystal of ionic solute M^+X^- is placed in water, initially the crystal just dissolves producing $M^+(aq)$ ions and $X^-(aq)$ ions. As the concentration of the ions in solution increases, however, the likelihood of oppositely charged ions colliding and *re*forming the solid increases. Eventually, an equilibrium is reached in which dissolving and reforming of the solid are occurring at the same speed. Past this point in time, there is no further net increase in the concentration of the dissolved ions.

44. The equilibrium represents the balancing of the dynamic processes of dissolving and of reformation of the solid. The amount of excess solid added in preparing a solution might affect the *speed* at which the point of equilibrium is reached, but it will not affect the net amount of solute present in solution at the point of equilibrium. A given amount of solvent can only "hold" a certain amount of solute.

45. Stirring or grinding the solute increases the speed with which the solute dissolves, but the ultimate *amount* of solute that dissolves is fixed by the equilibrium constant for the dissolving process, K_{sp}.

46. a. $NiS(s) \rightleftharpoons Ni^{2+}(aq) + S^{2-}(aq)$

$K_{sp} = [Ni^{2+}][S^{2-}]$

b. $CuCO_3(s) \rightleftharpoons Cu^{2+}(aq) + CO_3^{2-}(aq)$

$K_{sp} = [Cu^{2+}][CO_3^{2-}]$

c. $BaCrO_4(s) \rightleftharpoons Ba^{2+}(aq) + CrO_4^{2-}(aq)$

$K_{sp} = [Ba^{2+}][CrO_4^{2-}]$

d. $Ag_3PO_4(s) \rightleftharpoons 3Ag^+(aq) + PO_4^{3-}(aq)$

$K_{sp} = [Ag^+]^3[PO_4^{3-}]$

47. a. $PbBr_2(s) \rightleftharpoons Pb^{2+}(aq) + 2Br^-(aq)$

$K_{sp} = [Pb^{2+}][Br^-]^2$

b. $Ag_2S(s) \rightleftharpoons 2Ag^+(aq) + S^{2-}(aq)$

$K_{sp} = [Ag^+]^2[S^{2-}]$

c. $PbCO_3(s) \rightleftharpoons Pb^{2+}(aq) + CO_3^{2-}(aq)$

$K_{sp} = [Pb^{2+}][CO_3^{2-}]$

d. $Sr_3(PO_4)_2(s) \rightleftharpoons 3Sr^{2+}(aq) + 2PO_4^{3-}(aq)$

$K_{sp} = [Sr^{2+}]^3[PO_4^{3-}]^2$

48. $ZnCO_3(s) \rightleftharpoons Zn^{2+}(aq) + CO_3^{2-}(aq)$

molar mass $ZnCO_3 = 125.4$ g

$M = \dfrac{1.12 \times 10^{-4} \text{ g}}{1 \text{ L}} \times \dfrac{1 \text{ mol}}{125.4 \text{ g}} = 8.93 \times 10^{-7} M$

$K_{sp} = [Zn^{2+}][CO_3^{2-}] = (8.93 \times 10^{-7} M)(8.93 \times 10^{-7} M) = 7.98 \times 10^{-13}$

49. $NiS(s) \rightleftharpoons Ni^{2+}(aq) + S^{2-}(aq)$

molar mass $NiS = 90.77$ g

3.6×10^{-4} g NiS/L $\times \dfrac{1 \text{ mol NiS}}{90.77 \text{ g NiS}} = 3.97 \times 10^{-6} M$

$K_{sp} = [Ni^{2+}][S^{2-}] = (3.97 \times 10^{-6} M)(3.97 \times 10^{-6} M) = 1.6 \times 10^{-11}$

50. $CaCO_3(s) \rightleftharpoons Ca^{2+}(aq) + CO_3^{2-}(aq)$

$K_{sp} = [Ca^{2+}][CO_3^{2-}] = 3.0 \times 10^{-9}$

Let x represent the number of moles of $CaCO_3(s)$ that dissolve per liter, then $[Ca^{2+}] = x$ and $[CO_3^{2-}] = x$ also from the stoichiometry of the reaction; then

$K_{sp} = [x][x] = x^2 = 3.0 \times 10^{-9}$

$x = [CaCO_3] = 5.5 \times 10^{-5} M$ (5.5×10^{-3} g/L)

51. $PbCrO_4(s) \rightleftharpoons Pb^{2+}(aq) + CrO_4^{2-}(aq)$

$K_{sp} = [Pb^{2+}][CrO_4^{2-}] = 2.8 \times 10^{-13}$

Let x represent the number of moles of $PbCrO_4(s)$ that dissolve per liter. Then $[Pb^{2+}] = x$ and $[CrO_4^{2-}] = x$ also from the stoichiometry of the reaction; then

$K_{sp} = [x][x] = x^2 = 2.8 \times 10^{-13}$

$x = [PbCrO_4] = 5.3 \times 10^{-7} M$

Molar mass of $PbCrO_4 = 323.2$ g

5.3×10^{-7} mol/L $\times 323.2$ g/mol $= 1.7 \times 10^{-4}$ g/L

Equilibrium

52. $PbCl_2(s) \rightleftharpoons Pb^{2+}(aq) + 2Cl^-(aq)$

    $K_{sp} = [Pb^{2+}][Cl^-]^2$

    If $PbCl_2$ dissolves to the extent of 3.6×10^{-2} M, then

    $[Pb^{2+}] = 3.6 \times 10^{-2}$ M and

    $[Cl^-] = 2 \times (3.6 \times 10^{-2}) = 7.2 \times 10^{-2}$ M

    $K_{sp} = (3.6 \times 10^{-2} M)(7.2 \times 10^{-2} M)^2 = 1.9 \times 10^{-4}$

    molar mass $PbCl_2$ = 278.1 g

    $3.6 \times 10^{-2} \dfrac{mol}{1\ L} \times \dfrac{278.1\ g}{1\ mol} = 10.\ g/L$

53. $Hg_2Cl_2(s) \rightleftharpoons Hg_2^{2+}(aq) + 2Cl^-(aq)$

    $K_{sp} = [Hg_2^{2+}][Cl^-]^2$

    let x represent the number of moles of Hg_2Cl_2 that dissolve per liter; then

    $[Hg_2^{2+}] = x$ and $[Cl^-] = 2x$

    $K_{sp} = [x][2x]^2 = 4x^3 = 1.3 \times 10^{-18}$

    $x^3 = 3.25 \times 10^{-19}$

    $x = [Hg_2^{2+}] = 6.9 \times 10^{-7}$ M

54. Collision between molecules is *not* the only prerequisite for a reaction. The molecules must also possess sufficient energy to react with each other, and must have the proper spatial orientation for reaction.

55. An increase in temperature increases the fraction of molecules that possess sufficient energy for a collision to result in a reaction.

56. To say a reaction is *reversible* means that, to one extent or another, the reaction may occur in either direction.

57. When we say that a chemical equilibrium is *dynamic*, we are recognizing the fact that even though the reaction has appeared macroscopically to have stopped, on a microscopic basis the forward and reverse reactions are still taking place, at the same speed.

58. Heat is considered a *product* of an exothermic process. Adding a product to a system in equilibrium causes the reverse reaction to occur (producing additional reactants).

59. $PCl_5(g) \rightleftharpoons PCl_3(g) + Cl_2(g)$

    $K = \dfrac{[PCl_3][Cl_2]}{[PCl_5]} = 4.5 \times 10^{-3}$

    The concentration of PCl_5 is to be twice the concentration of PCl_3:

    $[PCl_5] = 2 \times [PCl_3]$

$$K = \frac{[PCl_3][Cl_2]}{2 \times [PCl_3]} = 4.5 \times 10^{-3}$$

$$K = \frac{[Cl_2]}{2} = 4.5 \times 10^{-3} \quad \text{and} \quad [Cl_2] = 9.0 \times 10^{-3}\ M$$

60. A higher concentration means there are more molecules present, which results in a greater frequency of collision between molecules.

61. At higher temperatures, the average kinetic energy of the reactant molecules is larger. At higher temperatures, the probability that a collision between molecules will be energetic enough for reaction to take place is larger. On a molecular basis, a higher temperature means a given molecule will be moving faster.

62. When a liquid is confined in an otherwise empty closed container, the liquid begins to evaporate, producing molecules of vapor in the empty space of the container. As the amount of vapor increases, molecules in the vapor phase begin to condense and reenter the liquid state. Eventually the opposite processes of evaporation and condensation will be going on at the same speed: beyond this point, for every molecule that leaves the liquid state and evaporates, there is a molecule of vapor which leaves the vapor state and condenses. We know the state of equilibrium has been reached when there is no further change in the pressure of the vapor.

63. a. $K = \dfrac{[HBr]^2}{[H_2][Br_2]}$

   b. $K = \dfrac{[H_2S]^2}{[H_2]^2[S_2]}$

   c. $K = \dfrac{[HCN]^2}{[H_2][C_2N_2]}$

64. a. $K = \dfrac{1}{[O_2]^3}$

   b. $K = \dfrac{1}{[NH_3][HCl]}$

   c. $K = \dfrac{1}{[O_2]}$

65. $2NO(g) + O_2(g) \rightleftharpoons 2NO_2(g)$

   a. shifts to right

   b. shifts to right

   c. no effect (He is not involved in the reaction)

66. molar mass Ni(OH)$_2$ = 92.71 g

$$\frac{0.14 \text{ g Ni(OH)}_2}{1.00 \text{ L}} \times \frac{1 \text{ mol Ni(OH)}_2}{92.71 \text{ g Ni(OH)}_2} = 1.510 \times 10^{-3} \, M$$

Ni(OH)$_2$(s) $\rightleftharpoons$ Ni^{2+}(aq) + 2OH$^-$(aq)

K_{sp} = [Ni^{2+}][OH$^-$]^2

If 1.510 x 10^{-3} M of Ni(OH)$_2$ dissolves, then

[Ni^{2+}] = 1.510 x 10^{-3} M and [OH$^-$] = 2 x (1.510 x 10^{-3} M) = 3.020 x 10^{-3} M

K_{sp} = (1.510 x 10^{-3} M)(3.020 x 10^{-3} M)^2 = 1.4 x 10^{-8}

Chapter 18 Oxidation-Reduction Reactions/Electrochemistry

1. Oxidation can be defined as the loss of electrons by an atom, molecule, or ion. Oxidation may also be defined as an increase in oxidation state for an element, but since elements can only increase their oxidation states by losing electrons, the two definitions are equivalent. The following equation shows the oxidation of copper metal to copper(II) ion.

    $$Cu(s) \rightarrow Cu^{2+}(aq) + 2e^-$$

2. Reduction can be defined as the gaining of electrons by an atom, molecule, or ion. Reduction may also be defined as a decrease in oxidation state for an element, but naturally such a decrease takes place by the gaining of electrons (so the two definitions are equivalent).

    $S + 2e^- \rightarrow S^{2-}$ is an example of a reduction process.

3. Each of these reactions involves a *metallic* element in the form of the free element on one side of the equation; on the other side of the equation, the metallic element is *combined* in an ionic compound. If a metallic element goes from the free metal to the ionic form, the metal is oxidized (loses electrons). Thus, the nonmetal is reduced.

    a. sodium is oxidized, nitrogen is reduced
    b. magnesium is oxidized, chlorine is reduced
    c. aluminum is oxidized, bromine is reduced
    d. iron is oxidized, oxygen is reduced

4. Most of these reactions involve a *metallic* element in the form of the *free* element on one side of the equation; on the other side of the equation, the metallic element is *combined* in an ionic compound. If a metallic element goes from the free metal to the ionic form, the metal is oxidized (loses electrons). Thus, the nonmetal is reduced.

    a. magnesium is oxidized, bromine is reduced
    b. sodium is oxidized, sulfur is reduced
    c. bromide ion is oxidized, chlorine is reduced
    d. potassium is oxidized, nitrogen is reduced

5. The assignment of oxidation states is a bookkeeping method by which *charges* are assigned to the various atoms in a compound: this method allows us to keep track of electrons transferred between species in oxidation-reduction reactions.

6. For the very electronegative elements (such as oxygen), we assign each of these elements an oxidation state equal to its charge when the element forms an anion. Since oxygen forms O^{2-} ions, we assign oxygen an oxidation state of –2 even in compounds where no ions exist. The most common situation in which oxygen is *not* assigned the –2 oxidation state occurs with the *peroxide* ion, O_2^{2-}, in which each oxygen atom is in the –1 oxidation state.

7. Oxidation states represent a bookkeeping method to assign electrons in a molecule or ion. Since a neutral molecule has an overall charge of zero, the sum of the oxidation states in

Oxidation-Reduction Reactions/Electrochemistry

a neutral molecule must be zero. For example, water (H_2O) is a neutral molecule with zero overall charge. Each hydrogen in water has an oxidation state of +1, whereas oxygen has an oxidation state of –2: $2(+1) + (-2) = 0$

8. Oxidation states represent a bookkeeping method to assign electrons in a molecule or ion. Since an ion has a net charge, the sum of the oxidation states of the atoms in the ion must equal the charge on the ion. For example, the hydroxide ion (OH^-) has an overall charge of –1 because hydrogen has an oxidation state of +1, whereas oxygen has an oxidation state of –2 in the hydroxide ion: $(-2) + (+1) = -1$

9. The rules for assigning oxidation states are given in the text. The rule which applies for each element in the following answers is given in parentheses after the element and its oxidation state.

   a. N, +3 (Rule 6); Cl, –1 (Rule 5)
   b. S, +6 (Rule 6); F, –1 (Rule 5)
   c. P, +5 (Rule 6); Cl, –1 (Rule 5)
   d. Si, –4 (Rule 6); H, +1 (Rule 4)

10. The rules for assigning oxidation states are given in the text. The rule which applies for each element in the following answers is given in parentheses after the element and its oxidation state.

    a. H, +1 (Rule 4); Br, –1 (Rule 6 or Rule 5)
    b. H, +1 (Rule 4); O, –2 (Rule 3); Br, +1 (Rule 6)
    c. Br, 0 (Rule 1)
    d. H, +1 (Rule 4); O, –2 (Rule 3); Br, +7 (Rule 6)

11. The rules for assigning oxidation states are given in the text. The rule which applies for each element in the following answers is given in parentheses after the element and its oxidation state.

    a. H, +1 (Rule 4); O, –2 (Rule 3); N, +5 (Rule 6)
    b. H, +1 (Rule 4); O, –2 (Rule 3); P, +5 (Rule 7)
    c. H, +1 (Rule 4); O, –2 (Rule 3); S, +6 (Rule 7)
    d. O, –1 (Rule 3)

12. The rules for assigning oxidation states are given in the text. The rule which applies for each element in the following answers is given in parentheses after the element and its oxidation state.

    a. Cu, +2 (Rules 2,6); Cl, –1 (Rules 2,5)
    b. Cr, +3 (Rules 2,6); Cl, –1 (Rules 2,5)
    c. H, +1 (Rule 4); O, –2 (Rule 3); Cr, +6 (Rule 7)
    d. Cr, +3 (Rule 2,6); O, –2 (Rules 2,3)

Copyright © Houghton Mifflin Company. All rights reserved.

13. The rules for assigning oxidation states are given in the text. The rule which applies for each element in the following answers is given in parentheses after the element and its oxidation state.

   a. H, +1 (Rule 4); C, −4 (Rule 6)

   b. Na, +1 (Rule 2); O, −2 (Rule 3); C, +4 (Rule 6)

   c. K, +1 (Rule 2); H, +1 (Rule 4); O, −2 (Rule 3); C, +4 (Rule 6)

   d. O, −2 (Rule 3); C, +2 (Rule 6)

14. Consider the following simple oxidation reaction

   $$Na \rightarrow Na^+ + e^-$$

   Clearly the sodium atom on the left side of the equation is losing an electron in forming the sodium ion on the right side of the equation. The sodium atom on the left side of the equation is in the *zero* oxidation state because it represents a pure element. The sodium ion on the right side of the equation is in the +1 oxidation state (the same as the charge on the simple ion). Thus, by losing one electron, sodium has increased in oxidation state by one unit.

15. Electrons are negative; when an atom gains electrons, it gains one negative charge for each electron gained. For example, in the reduction reaction $Cl + e^- \rightarrow Cl^-$, the oxidation state of chlorine decreases from 0 to −1 as the electron is gained.

16. An oxidizing agent *causes* another species to be oxidized (to lose electrons). In order to make another species lose electrons, the oxidizing agent must be capable of gaining the electrons; an oxidizing agent is itself reduced. On the contrary, a reducing agent is itself oxidized.

17. An oxidizing agent oxidizes another species by gaining the electrons lost by the other species; therefore, an oxidizing agent itself decreases in oxidation state. A reducing agent increases its oxidation state when acting on another atom or molecule.

18. a. $Zn(s) + 2HNO_3(g) \rightarrow Zn(NO_3)_2(aq) + H_2(g)$

   Zn 0   H +1   Zn +2   H 0
          N +5   N +5
          O −2   O −2

   zinc is oxidized; hydrogen is reduced

   b. $H_2(g) + CuSO_4(aq) \rightarrow Cu(s) + H_2SO_4(aq)$

   H 0   Cu +2   Cu 0   H +1
         S +6   S +6
         O −2   O −2

   hydrogen is oxidized; copper is reduced

Oxidation-Reduction Reactions/Electrochemistry

c. $N_2(g) + 3Br_2(l) \rightarrow 2NBr_3(g)$

N 0 Br 0 N −3

        Br +1

N is more electronegative than Br
nitrogen is reduced; bromine is oxidized

d. $2KBr(aq) + Cl_2(g) \rightarrow 2KCl(aq) + Br_2(l)$

K +1 Cl 0 K +1 Br 0

Br −1   Cl −1

bromide is oxidized; chlorine is reduced

19. a. $Cu(s) + 2AgNO_3(aq) \rightarrow 2Ag(s) + Cu(NO_3)_2(aq)$

Cu 0 Ag +1   Ag 0 Cu +2

    N +5   N +5

    O −2   O −2

copper is oxidized; silver is reduced

b. $N_2(g) + 3F_2(g) \rightarrow 2NF_3(g)$

N 0 F 0 F −1

       N +3

nitrogen is oxidized; fluorine is reduced

c. $2Fe_2O_3(s) + 3S(s) \rightarrow 4Fe(s) + 3SO_2(g)$

Fe +3 S 0 Fe 0 S +4
O −2   O −2

sulfur is oxidized; iron is reduced

d. $2H_2O_2(l) \rightarrow 2H_2O(l) + O_2(g)$

H +1 H +1 O 0
O −1 O −2

oxygen is both oxidized and reduced

20. Silver is reduced [+1 in AgBr(s), 0 in Ag(s)]; bromine is oxidized [−1 in AgBr, 0 in $Br_2(g)$].

21. Magnesium is oxidized [0 in Mg(s), +2 in $Mg(OH)_2(s)$]; hydrogen is reduced [+1 in $H_2O(l)$, 0 in $H_2(g)$].

22. Oxidation-reduction reactions must be balanced with respect to *mass* (the total number of each type of atom on each side of the balanced equation must be the same) and with respect to *charge* (whatever number of electrons is lost in the oxidation process must be gained in the reduction process, with no "extra" electrons).

23. Under ordinary conditions it is impossible to have "free" electrons that are not part of some atom, ion, or molecule. For this reason, the total number of electrons lost by the species being oxidized must equal the total number of electrons gained by the species being reduced.

24. a. $N_2(g) \rightarrow N^{3-}(aq)$

   Balance nitrogen: $N_2(g) \rightarrow \mathbf{2}N^{3-}(aq)$

   Balance charge: $\mathbf{6e^-} + N_2(g) \rightarrow 2N^{3-}(aq)$

   Balanced half reaction: $6e^- + N_2(g) \rightarrow 2N^{3-}(aq)$

   b. $O_2^{2-}(aq) \rightarrow O_2(g)$

   Balance charge: $O_2^{2-}(aq) \rightarrow O_2(g) + \mathbf{2e^-}$

   Balanced half reaction: $O_2^{2-}(aq) \rightarrow O_2(g) + 2e^-$

   c. $Zn(s) \rightarrow Zn^{2+}(aq)$

   Balance charge: $Zn(s) \rightarrow Zn^{2+}(aq) + \mathbf{2e^-}$

   Balanced half reaction: $Zn(s) \rightarrow Zn^{2+}(aq) + 2e^-$

   d. $F_2(g) \rightarrow F^-(aq)$

   Balance fluorine: $F_2(g) \rightarrow \mathbf{2}F^-(aq)$

   Balance charge: $\mathbf{2e^-} + F_2(g) \rightarrow 2F^-(aq)$

   Balanced half reaction: $2e^- + F_2(g) \rightarrow 2F^-(aq)$

25. a. $O_2(g) \rightarrow H_2O(l)$

   Balance oxygen: $O_2(g) \rightarrow \mathbf{2}H_2O(l)$

   Balance hydrogen: $\mathbf{4H^+}(aq) + O_2(g) \rightarrow 2H_2O(l)$

   Balance charge: $\mathbf{4e^-} + 4H^+(aq) + O_2(g) \rightarrow 2H_2O(l)$

   Balanced half reaction: $4e^- + 4H^+(aq) + O_2(g) \rightarrow 2H_2O(l)$

   b. $IO_3^-(aq) \rightarrow I_2(s)$

   Balance iodine: $\mathbf{2}IO_3^-(aq) \rightarrow I_2(s)$

   Balance oxygen: $2IO_3^-(aq) \rightarrow I_2(s) + \mathbf{6H_2O}(l)$

   Balance hydrogen: $\mathbf{12H^+} + 2IO_3^-(aq) \rightarrow I_2(s) + 6H_2O(l)$

   Balance charge: $12H^+(aq) + 2IO_3^-(aq) + \mathbf{10e^-} \rightarrow I_2(s) + 6H_2O(l)$

   Balanced half reaction: $12H^+(aq) + 2IO_3^-(aq) + 10e^- \rightarrow I_2(s) + 6H_2O(l)$

   c. $VO^{2+}(aq) \rightarrow V^{3+}(aq)$

   Balance oxygen: $VO^{2+}(aq) \rightarrow V^{3+}(aq) + \mathbf{H_2O}(l)$

Oxidation-Reduction Reactions/Electrochemistry

>   Balance hydrogen: $2H^+(aq) + VO^{2+}(aq) \rightarrow V^{3+}(aq) + H_2O(l)$
>
>   Balance charge: $e^- + 2H^+(aq) + VO^{2+}(aq) \rightarrow V^{3+}(aq) + H_2O(l)$
>
>   Balanced half reaction: $e^- + 2H^+(aq) + VO^{2+}(aq) \rightarrow V^{3+}(aq) + H_2O(l)$

  d. $BiO^+(aq) \rightarrow Bi(s)$

>   Balance oxygen: $BiO^+(aq) \rightarrow Bi(s) + H_2O(l)$
>
>   Balance hydrogen: $2H^+(aq) + BiO^+(aq) \rightarrow Bi(s) + H_2O(l)$
>
>   Balance charge: $3e^- + 2H^+(aq) + BiO^+(aq) \rightarrow Bi(s) + H_2O(l)$
>
>   Balanced half reaction: $3e^- + 2H^+(aq) + BiO^+(aq) \rightarrow Bi(s) + H_2O(l)$

26. For simplicity, the physical states of the substances have been omitted until the final balanced equation is given.

  a. $MnO_4^-(aq) + Zn(s) \rightarrow Mn^{2+}(aq) + Zn^{2+}(aq)$

> $MnO_4^- \rightarrow Mn^{2+}$
>
>   Balance oxygen: $MnO_4^- \rightarrow Mn^{2+} + 4H_2O$
>
>   Balance hydrogen: $8H^+ + MnO_4^- \rightarrow Mn^{2+} + 4H_2O$
>
>   Balance charge: $5e^- + 8H^+ + MnO_4^- \rightarrow Mn^{2+} + 4H_2O$
>
> $Zn \rightarrow Zn^{2+}$
>
>   Balance charge: $Zn \rightarrow Zn^{2+} + 2e^-$
>
>   Combine half reactions: $2 \times (5e^- + 8H^+ + MnO_4^- \rightarrow Mn^{2+} + 4H_2O)$
> $$5 \times (Zn \rightarrow Zn^{2+} + 2e^-)$$
>
> $16H^+(aq) + 2MnO_4^-(aq) + 5Zn(s) \rightarrow 2Mn^{2+}(aq) + 8H_2O(l) + 5Zn^{2+}(aq)$

  b. $Sn^{4+}(aq) + H_2(g) \rightarrow Sn^{2+}(aq) + H^+(aq)$

> $Sn^{4+} \rightarrow Sn^{2+}$
>
>   Balance charge: $Sn^{4+} + 2e^- \rightarrow Sn^{2+}$
>
> $H_2 \rightarrow H^+$
>
>   Balance hydrogen: $H_2 \rightarrow 2H^+$
>
>   Balance charge: $H_2 \rightarrow 2H^+ + 2e^-$
>
> $Sn^{4+}(aq) + H_2(g) \rightarrow Sn^{2+}(aq) + 2H^+(aq)$

  c. $Zn(s) + NO_3^-(aq) \rightarrow Zn^{2+}(aq) + NO_2(g)$

> $Zn \rightarrow Zn^{2+}$
>
>   Balance charge: $Zn \rightarrow Zn^{2+} + 2e^-$

$NO_3^- \rightarrow NO_2$

Balance oxygen: $NO_3^- \rightarrow NO_2 + \mathbf{H_2O}$

Balance hydrogen: $NO_3^- + \mathbf{2H^+} \rightarrow NO_2 + H_2O$

Balance charge: $NO_3^- + 2H^+ + e^- \rightarrow NO_2 + H_2O$

Balanced half-reaction: $NO_3^- + 2H^+ + e^- \rightarrow NO_2 + H_2O$

Combine half reactions: $Zn \rightarrow Zn^{2+} + 2e^-$
$2 \times (NO_3^- + 2H^+ + e^- \rightarrow NO_2 + H_2O)$

$Zn(s) + 2NO_3^-(aq) + 4H^+(aq) \rightarrow Zn^{2+}(aq) + 2NO_2(g) + 2H_2O(l)$

d. $H_2S(g) + Br_2(l) \rightarrow S(s) + Br^-(aq)$

$H_2S \rightarrow S$

Balance hydrogen: $H_2S \rightarrow S + \mathbf{2H^+}$

Balance charge: $H_2S \rightarrow S + 2H^+ + \mathbf{2e^-}$

Balanced half-reaction: $H_2S \rightarrow S + 2H^+ + 2e^-$

$Br_2 \rightarrow Br^-$

Balance bromine: $Br_2 \rightarrow \mathbf{2Br^-}$

Balance charge: $\mathbf{2e^-} + Br_2 \rightarrow 2Br^-$

Balanced half-reaction: $2e^- + Br_2 \rightarrow 2Br^-$

$H_2S(g) + Br_2(l) \rightarrow S(s) + 2H^+(aq) + 2Br^-(aq)$

27. For simplicity, the physical states of the substances have been omitted until the final balanced equation is given.

For the oxidation of iodide ion, I^-, in acidic solution, the half-reaction is always the *same*:

$I^- \rightarrow I_2$

Balance iodine: $\mathbf{2I^-} \rightarrow I_2$

Balance charge: $2I^- \rightarrow I_2 + \mathbf{2e^-}$

Balanced half-reaction: $2I^- \rightarrow I_2 + 2e^-$

a. $IO_3^- \rightarrow I_2$

Balance iodine: $\mathbf{2IO_3^-} \rightarrow I_2$

Balance oxygen: $2IO_3^- \rightarrow I_2 + \mathbf{6H_2O}$

Balance hydrogen: $2IO_3^- + \mathbf{12H^+} \rightarrow I_2 + 6H_2O$

Balance charge: $2IO_3^- + 12H^+ + \mathbf{10e^-} \rightarrow I_2 + 6H_2O$

Balanced half-reaction: $2IO_3^- + 12H^+ + 10e^- \rightarrow I_2 + 6H_2O$

Combine the half-reactions: $2IO_3^- + 12H^+ + 10e^- \rightarrow I_2 + 6H_2O$

$$5 \times (2I^- \rightarrow I_2 + 2e^-)$$

$2IO_3^-(aq) + 12H^+(aq) + 10I^-(aq) \rightarrow 6I_2(aq) + 6H_2O(l)$

$IO_3^-(aq) + 6H^+(aq) + 5I^-(aq) \rightarrow 3I_2(aq) + 3H_2O(l)$

b. $Cr_2O_7^{2-} \rightarrow Cr^{3+}$

Balance chromium: $Cr_2O_7^- \rightarrow 2Cr^{3+}$

Balance oxygen: $Cr_2O_7^{2-} \rightarrow Cr^{3+} + \mathbf{7H_2O}$

Balance hydrogen: $Cr_2O_7^{2-} + \mathbf{14H^+} \rightarrow 2Cr^{3+} + 7H_2O$

Balance charge: $Cr_2O_7^{2-} + 14H^+ + \mathbf{6e^-} \rightarrow 2Cr^{3+} + 7H_2O$

Balanced half-reaction: $Cr_2O_7^{2-} + 14H^+ + 6e^- \rightarrow 2Cr^{3+} + 7H_2O$

Combine the half-reactions: $3 \times (2I^- \rightarrow I_2 + 2e^-)$

$$Cr_2O_7^{2-} + 14H^+ + 6e^- \rightarrow 2Cr^{3+} + 7H_2O$$

$6I^-(aq) + Cr_2O_7^{2-}(aq) + 14H^+(aq) \rightarrow 3I_2(aq) + 2Cr^{3+}(aq) + 7H_2O(l)$

c. $Cu^{2+} \rightarrow CuI$

Balance iodine: $Cu^{2+} + I^- \rightarrow CuI$

Balance charge: $Cu^{2+} + I^- + e^- \rightarrow CuI$

Balanced half-reaction: $Cu^{2+} + I^- + e^- \rightarrow CuI$

Combine the half-reactions: $2I^- \rightarrow I_2 + 2e^-$

$$2 \times (Cu^{2+} + I^- + e^- \rightarrow CuI)$$

$2Cu^{2+}(aq) + 4I^-(aq) \rightarrow 2CuI(s) + I_2(aq)$

28. An oxidation-reduction is made useful as a galvanic cell (battery) by physically separating the oxidation half-reaction from the reduction half-reaction, and by causing the electrons to be transferred through a *wire* connecting the two. The passage of electrons through the wire represents an electrical current, which might be used to drive a motor or some other device.

29. A salt bridge typically consists of a *U*-shaped tube filled with an inert electrolyte (one involving ions that are not part of the oxidation-reduction reaction). A salt bridge is used to complete the electrical circuit in a cell. Any method which allows transfer of charge without allowing bulk mixing of the solutions may be used (another common method is to set up one half-cell in a porous cup, which is then placed in the beaker containing the second half-cell).

30. In a galvanic cell, electrons flow from the anode (where oxidation occurs) to the cathode (where reduction occurs).

31.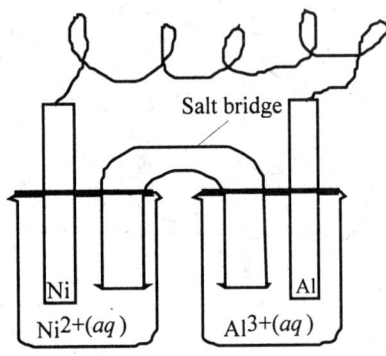

Ni^{2+}(aq) ion is reduced; Al(s) is oxidized.

The reaction at the anode is Al(s) → Al^{3+}(aq) + 3e$^-$.

The reaction at the cathode is Ni^{2+}(aq) + 2e$^-$ → Ni(s).

32.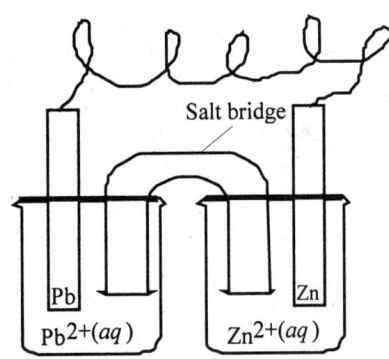

Pb^{2+}(aq) ion is reduced; Zn(s) is oxidized.

The reaction at the anode is Zn(s) → Zn^{2+}(aq) + 2e$^-$.

The reaction at the cathode is Pb^{2+}(aq) + 2e$^-$ → Pb(s).

33. The overall reaction is

$$Pb(s) + PbO_2(s) + 2H_2SO_4(aq) \rightarrow 2PbSO_4(s) + 2H_2O(l)$$

in which Pb^0(s) is oxidized to Pb^{2+} and Pb^{IV}O$_2$ is reduced to Pb^{2+}. This reaction can be reversed by *electrolysis* of the mixture of water and PbSO$_4$(s) (passing electrical energy into the mixture from the outside).

34. Both normal and alkaline cells contain zinc as one electrode; zinc corrodes more slowly under alkaline conditions than in the highly acidic environment of a normal dry cell.

anode: Zn(s) + 2OH$^-$(aq) → ZnO(s) + H$_2$O(l) + 2e$^-$

cathode: 2MnO$_2$(s) + H$_2$O(l) + 2e$^-$ → Mn$_2$O$_3$(s) + 2OH$^-$(aq)

35. Corrosion represents returning metals to the natural state (ore) and involves *oxidation* of the metal. Corrosion of a metal is undesirable because, as the metal is converted to its oxide, the bulk of the metal loses its strength, flexibility, and other metallic properties. If the metal were part of some constructed item, the item would slowly disintegrate.

36. Aluminum is a very reactive metal when freshly isolated in the pure state. However, on standing for even a relatively short period of time, aluminum metal forms a thin coating of Al_2O_3 on its surface from reaction with atmospheric oxygen. This coating of Al_2O_3 is much less reactive than the metal and serves to protect the surface of the metal from further attack.

37. Most steels contain additives such as chromium or nickel. These additives are able to form protective oxide coatings on the surface of the steel which tend to prevent further oxidation.

38. In cathodic protection of steel tanks and pipes, a more reactive metal than iron is connected to the item to be protected. The active metal is then preferentially oxidized rather than the iron of the tank or pipe.

39. Electrolysis is the process of forcing an electrical current through a cell to produce a chemical change that would otherwise not occur on its own.

40. The main recharging reaction for the lead storage battery is

    $2PbSO_4(s) + 2H_2O(l) \rightarrow Pb(s) + PbO_2(s) + 2H_2SO_4(aq)$

    A major side reaction is the electrolysis of water

    $2H_2O(l) \rightarrow 2H_2(g) + O_2(g)$

    resulting in the production of an explosive mixture of hydrogen and oxygen, which accounts for many accidents during the recharging of such batteries.

41. Many metals can be produced by electrolysis of aqueous solutions of their salts, or by electrolysis of molten salts (for those metals that would react with water). Aluminum combines so readily with oxygen that it cannot be prepared by either of these methods. Aluminum is most commonly prepared by the Hall process, which is the electrolysis of a molten mixture of Al_2O_3 and Na_3AlF_6.

42. Electrolysis is applied in electroplating by making the item to be plated the cathode in a cell containing a solution of ions of the desired plating metal.

43. oxidation, reduced

44. lose

45. hydrogen; oxygen

46. a. $4Fe(s) + 3O_2(g) \rightarrow 2Fe_2O_3(s)$

    iron is oxidized, oxygen is reduced

b. $2Al(s) + 3Cl_2(g) \rightarrow 2AlCl_3(s)$

aluminum is oxidized, chlorine is reduced

c. $6Mg(s) + P_4(s) \rightarrow 2Mg_3P_2(s)$

magnesium is oxidized, phosphorus is reduced

47.
a. $2MnO_4^-(aq) + 6H^+(aq) + 5H_2O_2(aq) \rightarrow 2Mn^{2+}(aq) + 8H_2O(l) + 5O_2(g)$

b. $6Cu^+(aq) + 6H^+(aq) + BrO_3^-(aq) \rightarrow 6Cu^{2+}(aq) + Br^-(aq) + 3H_2O(l)$

c. $2HNO_2(aq) + 2H^+(aq) + 2I^-(aq) \rightarrow 2NO(g) + I_2(aq) + 2H_2O(l)$

48. Each of these reactions involves a *metallic* element in the form of the *free* element on one side of the equation; on the other side of the equation, the metallic element is *combined* in an ionic compound. If a metallic element goes from the free metal to the ionic form, the metal is oxidized (loses electrons).

a.   zinc is oxidized, nitrogen is reduced

b.   cobalt is oxidized, sulfur is reduced

c.   potassium is oxidized, oxygen is reduced

d.   silver is oxidized, oxygen is reduced

49. The rules for assigning oxidation states are given in the text. The rule which applies for each element in the following answers is given in parentheses after the element and its oxidation state.

a.   Mn +4 (Rule 6); O –2 (Rule 3)

b.   Ba +2 (Rule 2); Cr +6 (Rule 6); O –2 (Rule 3)

c.   H +1 (Rule 4); S +4 (Rule 6); O –2 (Rule 3)

d.   Ca +2 (Rule 2); P +5 (Rule 6); O –2 (Rule 3)

50.
a. $2B_2O_3(s) + 6Cl_2(g) \rightarrow 4BCl_3(l) + 3O_2(g)$

B +3   Cl 0   B +3   O 0

O –2   Cl –1

oxygen is oxidized; chlorine is reduced

b. $GeH_4(g) + O_2(g) \rightarrow Ge(s) + 2H_2O(g)$

Ge –4   O 0   Ge 0   H +1

H +1   O –2

germanium is oxidized; oxygen is reduced

c. $C_2H_4(g) + Cl_2(g) \rightarrow C_2H_4Cl_2(l)$

      C −2   Cl 0   C −1

      H +1   H +1; Cl −1

carbon is oxidized; chlorine is reduced

d. $O_2(g) + 2F_2(g) \rightarrow 2OF_2(g)$

      O 0   F 0   O +2

                     F −1

oxygen is oxidized; fluorine is reduced

51.

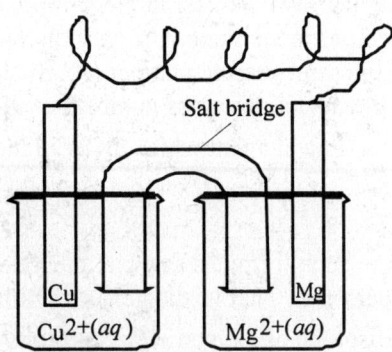

$Cu^{2+}(aq)$ ion is reduced; $Mg(s)$ is oxidized.

The reaction at the anode is $Mg(s) \rightarrow Mg^{2+}(aq) + 2e^-$.

The reaction at the cathode is $Cu^{2+}(aq) + 2e^- \rightarrow Cu(s)$.

Chapter 19   Radioactivity and Nuclear Energy

1. The nucleus of an atom has little or *no* effect on the atom's chemical properties. The chemical properties of an atom are determined by the number and arrangement of the atom's electrons (which are outside the nucleus).

2. The radius of a typical atomic nucleus is on the order of 10^{-13} cm, which is about one hundred thousand times smaller than the radius of an atom overall.

3. The atomic number (Z) of a nucleus represents the number of protons present in the nucleus. The mass number (A) of a nucleus represents the total number of protons and neutrons in the nucleus. For example, for the nuclide $^{13}_{6}C$, with six protons and seven neutrons, we have $Z = 6$ and $A = 13$.

4. Isotopes are atoms of the same element which differ in the number of neutrons present in the nuclei. Since the number of protons is the same in each such nucleus (the atoms are of the same element), isotopes have the same atomic number (Z). Since the number of neutrons present differs between the nuclei, the mass numbers of the nuclei (A) are different. Isotopes are atoms of the same element which differ in mass. For example, $^{13}_{6}C$ and $^{14}_{6}C$ both represent carbon atoms. However, $^{13}_{6}C$ one less neutron than does $^{14}_{6}C$.

5. The atomic number (Z) is written in such formulas as a left subscript, while the mass number (A) is written as a left superscript. That is, the general symbol for a nuclide is $^{A}_{Z}X$. As an example, consider the isotope of oxygen with 8 protons and 8 neutrons: its symbol would be $^{16}_{8}O$.

6. When a nucleus produces an alpha (α) particle, the atomic number of the nucleus decreases by two units.

7. When a nucleus produces a beta (β) particle, the atomic number of the parent nucleus is *increased* by *one* unit. The beta particle has a mass number of zero, but an "atomic number" of -1.

8. Gamma rays are high energy photons of electromagnetic radiation. Gamma rays are not considered to be particles. When a nucleus produces only gamma radiation, the atomic number and mass number of the nucleus do not change. Gamma rays represent the energy changes associated with transitions and rearrangement of the particles within the nucleus.

9. A positron is a particle with the same mass as an electron, but with the opposite charge: a positron is *positively* charged. It's mass number is therefore zero, and its "atomic number" is $+1$. When an unstable nucleus produces a positron, the mass number of the original nucleus is unchanged, but the atomic number of the original nucleus decreases by one unit.

10. Electron capture occurs when one of the inner orbital electrons is pulled into and becomes part of the nucleus.

Radioactivity and Nuclear Energy

11. $^{24}_{12}$Mg (12 protons, 12 neutrons)

    $^{25}_{12}$Mg (12 protons, 13 neutrons)

    $^{26}_{12}$Mg (12 protons, 14 neutrons)

12. a. $^{192}_{83}$Bi

    b. $^{204}_{82}$Pb

    c. $^{206}_{84}$Po

13. a. $^{0}_{-1}$e

    b. $^{0}_{-1}$e

    c. $^{210}_{83}$Bi

14. a. $^{136}_{53}$I → $^{136}_{54}$Xe + $^{0}_{-1}$e

    b. $^{133}_{51}$Sb → $^{133}_{52}$Te + $^{0}_{-1}$e

    c. $^{117}_{49}$I → $^{117}_{50}$Sn + $^{0}_{-1}$e

15. a. $^{226}_{88}$Ra → $^{222}_{86}$Rn + $^{4}_{2}$He

    b. $^{222}_{86}$Rn → $^{218}_{84}$Po + $^{4}_{2}$He

    c. $^{239}_{94}$Pu → $^{235}_{92}$U + $^{4}_{2}$He

    d. $^{8}_{4}$Be → $^{4}_{2}$He + $^{4}_{2}$He

16. A nuclear transformation represents the change of one element into another. Nuclear transformations are generally accomplished by bombardment of a target nucleus with some small, energetic particle that effects the desired transformation of the target nucleus.

17. There is often considerable repulsion between the target nucleus and the particles being used for bombardment (especially if the bombarding particle is positively charged like the target nucleus). Using accelerators to greatly speed up the bombarding particles can overcome this repulsion.

18. The elements with atomic number greater than 92 are referred to as the *transuranium* elements. The *transuranium* elements have been prepared by bombardment reactions of other nuclei.

19. $^{27}_{13}$Al + $^{4}_{2}$He → $^{30}_{15}$P + $^{1}_{0}$n

20. Geiger (Geiger-Müller) counters contain a probe which contains argon gas. The argon atoms themselves have no charge, but they can be ionized by high-energy particles from a radioactive decay process. Although a sample of normal uncharged argon gas does not

conduct an electrical current, argon gas that has been ionized will briefly conduct an electrical current (until the argon ions and electrons recombine). If an electric field is applied to the argon gas probe, then a brief pulse of electricity will be passed through the argon every time an ionization event occurs (every time a high-energy particle strikes the argon gas probe). The Geiger counter detects each pulse of current, and these pulses are then counted and displayed on the meter of the device. A scintillation counter uses a substance like sodium iodide, which emits light when struck by a high-energy particle from a radioactive decay. A detector senses the flashes of light from the sodium iodide, and these flashes are then counted and displayed on the meter of the device.

21. The half-life of a nucleus is the time required for one-half of the original sample of nuclei to decay. A given isotope of an element always has the same half-life, although different isotopes of the same element may have greatly different half-lives. Nuclei of different elements typically have different half-lives.

22. When we say that one nucleus is "hotter" than another, we mean that the "hot" nucleus undergoes more decay events per time period. The "hotness" of radionuclei is most commonly indicated by their *half-lives* (the amount of time required for half a sample to undergo the decay process). A nucleus with a short half-life will undergo more decay events in a given time than a nucleus with a long half-life.

23. highest   lowest
$^{24}Na > {}^{131}I > {}^{60}Co > {}^{3}H > {}^{14}C$

24. highest activity   lowest activity
$^{87}Sr > {}^{99}Tc > {}^{24}Na > {}^{99}Mo > {}^{133}Xe > {}^{131}I > {}^{32}P > {}^{51}Cr > {}^{59}Fe$

25. For ^{223}Ra, the half-life is 12 days. After two half-lives (24 days), 250 mg remains; after three half-lives (36 days), 125 mg remains.

    For ^{224}Ra, the half-life is 3.6 days. One month would be approximately 8 half-life periods (29 days), and approximately 4 mg remains.

    For ^{225}Ra, the half-life is 15 days. One month would be two half-life periods, and 250 mg remains.

26. For an administered dose of 100 μg, 0.39 μg remains after 2 days. The fraction remaining is 0.39/100 = 0.0039; on a percentage basis, less than 0.4% of the original radioisotope remains.

27. Carbon–14 ($^{14}_{6}C$) is most commonly used in the radiodating of archaeological artifacts.

28. Carbon–14 is produced in the upper atmosphere by the bombardment of ordinary nitrogen with neutrons from space:

    $^{14}_{7}N + {}^{1}_{0}n \rightarrow {}^{14}_{6}C + {}^{1}_{1}H$

29. The quantity of $^{14}_{6}C$ in the atmosphere is assumed to remain constant because a balance exists between the continued *production* of $^{14}_{6}C$ from bombardment of nitrogen in the upper atmosphere by cosmic rays, and the *decay* of $^{14}_{6}C$ through beta particle production.

30. We assume that the concentration of C–14 in the atmosphere is effectively constant. A living organism is constantly replenishing C–14 either through the processes of metabolism (sugars ingested in foods contain C–14), or photosynthesis (carbon dioxide contains C–14). When a plant dies, it no longer replenishes itself with C–14 from the atmosphere, and as the C–14 undergoes radioactive decay, its amount decreases with time.

31. A *radiotracer* is a radioactive nuclide that can be introduced into an organism in food or drugs, whose pathway through the body can then be traced by monitoring of the radioactivity of the nuclide. Carbon–14 and Phosphorus–32 have been used to study the conversion of nutrients into energy by living cells.

32. These isotopes and their uses are listed in Table 19.4. Some important examples include the use of I–131 (and other iodine isotopes) in the diagnosis and treatment of thyroid disease (iodine is used in the body primarily in the thyroid gland); Fe–59 in the study of the function of red blood cells (iron is a constituent of hemoglobin which is found in the red blood cells); Sr–87 in the study of bones (Sr is a Group 2 element, and is able to take the place of Ca in bone structures).

33. The forces that hold protons and neutrons together in the nucleus are *much greater* than the forces that bind atoms together in molecules.

34. Combining two light nuclei to form a heavier, more stable nucleus is called nuclear *fusion*. Splitting a heavy nucleus into nuclei with smaller mass numbers is called nuclear *fission*.

35. The energies released by nuclear processes are on the order of 10^6 times more powerful than those associated with ordinary chemical reactions.

36. $^{1}_{0}n + ^{235}_{92}U \rightarrow ^{142}_{56}Ba + ^{91}_{36}Kr + 3\,^{1}_{0}n$ is one possibility.

37. The fission of $^{235}_{92}U$ is initiated by bombardment of the nucleus with neutrons from an outside source. For every nucleus of $^{235}_{92}U$ that decays, however, three neutrons are produced by the process. Once the reaction has been started with neutrons from an outside source, the neutrons generated by the reaction itself can go on to cause other nuclei of $^{235}_{92}U$ to decay, thereby producing still more neutrons, and so on. All that is needed to sustain such a chain reaction is a sufficient density of $^{235}_{92}U$, so that the emitted neutrons are not lost to the outside.

38. A critical mass of a fissionable material is the amount needed to provide a high enough internal neutron flux to sustain the chain reaction (enough neutrons are produced to cause the continuous fission of further material). A sample with less than a critical mass is still radioactive, but cannot sustain a chain reaction.

39. The *moderator* in a uranium fission reactor surrounds the fuel rods and slows down the neutrons produced by the uranium decay process so that they can be absorbed more easily by other uranium atoms. The *control rods* are constructed of substances that absorb neutrons, and can be inserted into the reactor core to control the power level of the reactor. The *containment* of a reactor refers to the building in which the reactor core is located, which is designed to contain the radioactive core in the event of a nuclear accident. A *cooling liquid* (usually water) is circulated through the reactor to draw off the heat energy produced by the nuclear reaction, so that this heat energy can be converted to electrical energy in the power plant's turbines.

40. An actual nuclear explosion, of the type produced by a nuclear weapon, cannot occur in a nuclear reactor because the concentration of the fissionable materials is not sufficient to form a supercritical mass. However, since many reactors are cooled by water, which can decompose into hydrogen and oxygen gases, a *chemical* explosion is possible which could scatter the radioactive material used in the reactor.

41. If the system used to cool a reactor core fails, the reactor may reach temperatures high enough to melt the core itself. In a scenario referred to as the "China Syndrome," the molten reactor core could become hot enough so as to melt through the bottom of the reactor building and into the earth itself (eventually the molten material would reach cool ground water and resolidify, with possible release of radioactivity). If water is used to cool the reactor core, and the cooling system becomes blocked, it is possible for the heat from the reactor to cause a steam explosion (which would also release radioactivity), or to break down the coolant water into hydrogen gas (which could also explode).

42. $^{238}_{92}U + ^{1}_{0}n \rightarrow ^{239}_{92}U$

    $^{239}_{92}U \rightarrow ^{239}_{93}Np + ^{0}_{-1}e$

    $^{239}_{93}Np \rightarrow ^{239}_{94}Pu + ^{0}_{-1}e$

43. Nuclear *fusion* is the process of combining two light nuclei into a larger nucleus, with an energy release larger even than that provided by fission processes.

44. In one type of fusion reactor, two $^{2}_{1}H$ atoms are fused to produce $^{4}_{2}He$. Because the hydrogen nuclei are positively charged, extremely high energies (temperatures of 40 million K) are needed to overcome the repulsion between the nuclei as they are shot into each other.

45. Fusion produces an enormous amount of energy per gram of fused material. If it is hydrogen that is to be fused, the earth possesses an enormous supply of the needed raw materials in the oceans; the product nuclei from the fusion of hydrogen (helium isotopes) are far less dangerous than those produced by fission processes.

46. Although the energy transferred per event to a living creature is small, the quantity of energy is enough to break chemical bonds which may cause malfunctioning of cellular systems. In particular, many biochemical processes are chain-like in nature, and the production of a single odd ion in a cell by a radioactive event may have a cumulative effect. For example, ionization of a single bond in a sex cell may cause a drastic mutation in the creature resulting.

47. Somatic damage is damage directly to the organism itself, causing nearly immediate sickness or death to the organism. Genetic damage is damage to the genetic machinery of the organism, which will be manifested in future generations of offspring.

48. Alpha particles are stopped by the outermost layers of skin; beta particles penetrate only about 1 cm into the body; gamma rays are deeply penetrating.

49. Gamma rays penetrate long distances, but seldom cause ionization of biological molecules. Alpha particles, because they are much heavier although less penetrating, are very effective at ionizing biological molecules and leave a dense trail of damage in the organism. Isotopes which release alpha particles can be ingested or breathed into the body where the damage from the alpha particles will be more acute.

50. Nuclei of atoms that are chemically inert, or which are not ordinarily found in the body, tend to be excreted from the body quickly and do little damage. Other nuclei of atoms which form a part of the body's structure or normal metabolic processes are likely to be incorporated into the body. When a radioactive nuclide is ingested into the body, its capacity to cause damage also depends on how long it remains in the body. If the nuclide has been incorporated into the body, the danger is greatest.

51. The exposure limits given in Table 19.5 as causing no detectable clinical effect are 0 – 25 rem. The total yearly exposures from natural and human-induced radioactive sources are estimated in Table 19.6 as less than 200 *milli*rem (0.2 rem), which is well within the acceptable limits.

52. The decay series, in order from the top right of the diagram, is:

    alpha, beta, beta, alpha, alpha, alpha, alpha, alpha, beta, beta, alpha, beta, beta, alpha.

    This decay is indicated in color in the figure.

53. a. cobalt is a component of Vitamin B-12

    b. bones consist partly of $Ca_3(PO_4)_2$

    c. red blood cells contain hemoglobin, an iron-protein compound

    d. mercury is absorbed by substances in the brain (this is part of the reason mercury is so hazardous in the environment)

54. Despite the fact that nuclear waste has been generated for over 40 years, no permanent disposal plan has been implemented as yet. One proposal to dispose of such waste calls for the waste to be sealed in blocks of glass, which in turn are sealed in corrosion-proof metal drums, which would then be buried in deep, stable rock formations away from earthquake and other geologically active zones. In these deep storage areas, it is hoped that the waste could decay safely undisturbed until the radioactivity drops to "safe" levels.

55. $^{27}_{13}Al$: 13 protons, 14 neutrons

    $^{28}_{13}Al$: 13 protons, 15 neutrons

    $^{29}_{13}Al$: 13 protons, 16 neutrons

56. ^{131}I is used in the diagnosis and treatment of thyroid cancer and other dysfunctions of the thyroid gland. The thyroid gland is the only place in the human body which uses and stores iodine. I–131 that is administered concentrates in the thyroid, and can be used to cause an image on a scanner or x-ray film, or in higher doses, to selectively kill cancer cells in the thyroid. ^{201}Tl concentrates in healthy muscle cells when administered, and can be used to detect and assess damage to heart muscles after a heart attack: the damaged muscles show a lower uptake of Tl–201 than normal muscles.

57. Breeder reactors are set up to convert non-fissionable ^{238}U into fissionable ^{239}Pu. The material used for fission in a breeder reactor is a combination of U–235 (which undergoes fission in a chain reaction) and the more common U–238 isotope. Excess neutrons from the U–235 fission are absorbed by the U–238 converting it to the fissionable plutonium isotope Pu–239. Although Pu–239 is fissionable, its chemical and physical properties make it very difficult and expensive to handle and process.

Chapter 20   Organic Chemistry

1. Carbon has the unusual ability of bonding strongly to itself, forming long chains or rings of carbon atoms. Since there are many different possible arrangements for a long chain of carbon atoms, there exists a great multitude of possible carbon compounds.

2. A given carbon atom can be attached to a maximum of four other atoms. Carbon atoms have four valence electrons. By making four bonds, carbon atoms exactly complete their valence octet.

3. A double bond represents the sharing of an extra pair of electrons between two bonded atoms. Two examples are

$$\begin{array}{c} H \quad H \\ | \quad | \\ C = C \\ | \quad | \\ H \quad H \end{array} \qquad \begin{array}{c} H \quad H \quad H \\ | \quad | \quad | \\ C = C - C - H \\ | \quad \quad | \\ H \quad \quad H \end{array}$$

4. Each carbon atom in ethane is bonded to four other atoms. According to VSEPR theory, each carbon atom has its electron pairs arranged tetrahedrally.

5. A saturated hydrocarbon is one in which all the carbon-carbon bonds are single bonds, with each carbon atom forming bonds to four other atoms. The saturated hydrocarbons are called alkanes.

6. A "straight-chain" alkane is not really straight because the electron pairs on the carbon atoms have a tetrahedral orientation, separated by an angle of 109.5°. In order to give a truly straight chain, the angle between electron pairs would have to be 90° and multiples of 90°.

7. a.

$$\begin{array}{c} H \quad H \quad H \quad H \quad H \quad H \quad H \\ | \quad | \quad | \quad | \quad | \quad | \quad | \\ H-C-C-C-C-C-C-C-H \\ | \quad | \quad | \quad | \quad | \quad | \quad | \\ H \quad H \quad H \quad H \quad H \quad H \quad H \end{array}$$

heptane $CH_3CH_2CH_2CH_2CH_2CH_2CH_3$

b.

$$\begin{array}{c} H \quad H \quad H \quad H \quad H \quad H \quad H \quad H \quad H \\ | \quad | \quad | \quad | \quad | \quad | \quad | \quad | \quad | \\ H-C-C-C-C-C-C-C-C-C-H \\ | \quad | \quad | \quad | \quad | \quad | \quad | \quad | \quad | \\ H \quad H \quad H \quad H \quad H \quad H \quad H \quad H \quad H \end{array}$$

nonane $CH_3CH_2CH_2CH_2CH_2CH_2CH_2CH_2CH_3$

c.

```
     H   H   H
     |   |   |
 H — C — C — C — H
     |   |   |
     H   H   H
```
propane CH₃CH₂CH₃

d.

```
     H   H   H   H   H   H   H   H   H   H
     |   |   |   |   |   |   |   |   |   |
 H — C — C — C — C — C — C — C — C — C — C — H
     |   |   |   |   |   |   |   |   |   |
     H   H   H   H   H   H   H   H   H   H
```
decane CH₃CH₂CH₂CH₂CH₂CH₂CH₂CH₂CH₂CH₃

8. a.

```
     H   H   H   H
     |   |   |   |
 H — C — C — C — C — H
     |   |   |   |
     H   H   H   H
```
butane CH₃CH₂CH₂CH₃

b.

```
     H   H   H   H   H   H   H   H
     |   |   |   |   |   |   |   |
 H — C — C — C — C — C — C — C — C — H
     |   |   |   |   |   |   |   |
     H   H   H   H   H   H   H   H
```
octane CH₃CH₂CH₂CH₂CH₂CH₂CH₂CH₃

c.

```
     H   H   H   H   H
     |   |   |   |   |
 H — C — C — C — C — C — H
     |   |   |   |   |
     H   H   H   H   H
```
pentane CH₃CH₂CH₂CH₂CH₃

Organic Chemistry

d.

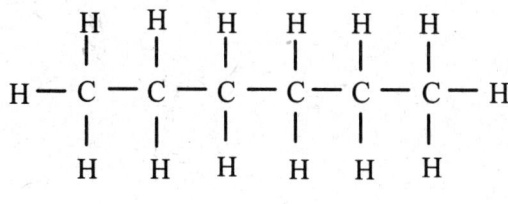

hexane CH₃CH₂CH₂CH₂CH₂CH₃

9. Structural isomerism occurs when two molecules have the same atoms present, but those atoms are bonded differently. The molecules have the same formulas but different arrangements of the atoms. The alkane butane is the first alkane to have an isomer:

$$\text{CH}_3\text{CH}_2\text{CH}_2\text{CH}_3 \qquad \begin{array}{c} \text{CH}_3 \\ | \\ \text{CH}_3\text{CHCH}_3 \end{array}$$

butane 　　　　　2-methylpropane

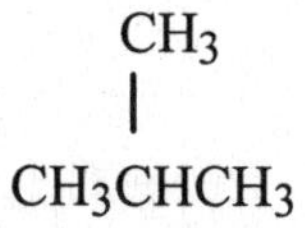

10. branch or substituent

11. With six carbon atoms, there are five isomers possible. Here are the carbon skeletons:

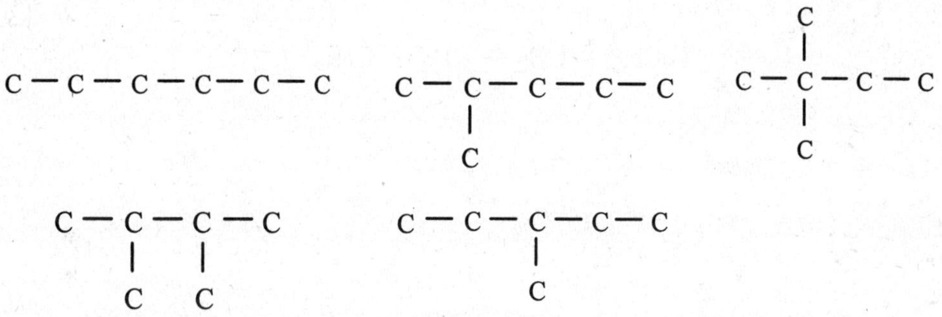

```
C—C—C—C—C—C      C—C—C—C—C        C—C—C—C
                      |                |
                      C                C
                                       C

C—C—C—C           C—C—C—C—C
  |  |              |
  C  C              C
```

12.
Number of carbons	root name
5	pentane (*pent-*)
6	hexane (*hex-*)
7	heptane (*hept-*)
8	octane (*oct-*)
9	nonane (*non-*)
10	decane (*dec-*)

13. The root name is derived from the number of carbon atoms in the *longest continuous* chain of carbon atoms.

14. The position of a substituent is indicated by a number which corresponds to the carbon atom in the longest chain to which the substituent is attached.

15. Multiple substituents are listed in alphabetical order, disregarding any prefix.

16.
   a. 3-ethylpentane
   b. 2,2-dimethylbutane
   c. 2,2-dimethylpropane
   d. 2,3,4-trimethylpentane

17.
   a.
$$CH_3-CH-CH_2-CH_2-CH_2-CH_3$$
$$|$$
$$CH_3$$

   b.
$$CH_3-CH_2-CH-CH_2-CH_2-CH_3$$
$$|$$
$$CH_3$$

   c.
$$CH_3$$
$$|$$
$$CH_3-C-CH_2-CH_2-CH_2-CH_3$$
$$|$$
$$CH_3$$

   d.
$$CH_3$$
$$|$$
$$CH_3-CH-CH-CH_2-CH_2-CH_3$$
$$|$$
$$CH_3$$

   e.
$$CH_3$$
$$|$$
$$CH_3-CH_2-C-CH_2-CH_2-CH_3$$
$$|$$
$$CH_3$$

18. Petroleum is a thick, dark liquid composed largely of hydrocarbons containing from 5 to more than 25 carbon atoms. Natural gas consists mostly of methane, but also may contain significant amounts of ethane, propane, and butane. These substances were formed over the eons from the decay of living organisms.

Organic Chemistry

19.
Number of C atoms	Use
C_5–C_{12}	gasoline
C_{10}–C_{18}	kerosene, jet fuel
C_{15}–C_{25}	diesel fuel, heating oil, lubrication
C_{25}–	asphalt

20. In the pyrolytic cracking of petroleum, the more abundant kerosene fraction of petroleum is heated to about 700°C, which causes the large molecules of the kerosene fraction to break into the smaller molecules characteristic of the gasoline fraction.

21. Tetraethyl lead was added to gasolines to prevent "knocking" of high efficiency automobile engines. The use of tetraethyl lead is being phased out because of the danger to the environment of the lead in this substance.

22. Alkanes are relatively unreactive because the C–C and C–H bonds that characterize these substances are relatively strong and difficult to break.

23. Combustion represents the vigorous reaction of a hydrocarbon (or other substance) with oxygen. The combustion of alkanes has been made use of as a source of heat, light, and mechanical energy.

24. In a substitution reaction, one or more of the hydrogen atoms of an alkane is *replaced* by another type of atom. Because of the unreactivity of alkanes, only very vigorous reactants (such as the halogens) are able to substitute for the hydrogen atoms of alkanes.

25. Dehydrogenation reactions involve the removal of hydrogen atoms from adjacent carbon atoms in an alkane (or other substance). When two hydrogen atoms are removed from an alkane or related compound, a double bond is created.

26. a. $2C_6H_{14}(l) + 19O_2(g) \rightarrow 12CO_2(g) + 14H_2O(g)$

   b. $CH_4(g) + Cl_2(g) \rightarrow CH_3Cl(l) + HCl(g)$

   c. $CHCl_3(l) + Cl_2(g) \rightarrow CCl_4(l) + HCl(g)$

27. Alkenes are hydrocarbons which contain a carbon–carbon double bond.

$$\diagdown \!\!\! \diagup \atop \diagup C = C \diagdown$$

The general formula for alkenes is C_nH_{2n} where n is the number of carbon atoms present.

28. An alkyne is a hydrocarbon containing a carbon-carbon triple bond. The general formula is C_nH_{2n-2}.

29.   a. $CH\equiv C-CH_3\,(g) + 2H_2\,(g) \longrightarrow CH_3-CH_2-CH_3\,(g)$

   b. $CH_3-CH=CH-CH_3\,(l) + Br_2\,(l) \longrightarrow CH_3-\underset{Br}{\underset{|}{CH}}-\underset{Br}{\underset{|}{CH}}-CH_3\,(l)$

   c. $2CH_3-C\equiv C-CH_3\,(l) + 11O_2\,(g) \longrightarrow 8CO_2\,(g) + 6H_2O\,(g)$

30.   a.   2-butene

   b.   3-methyl-1-butene

   c.   1-butyne

   d.   3-chloro-1-butene

31. The most obvious choices would be the normal alkenes with seven carbon atoms:

   $CH_2=CH-CH_2-CH_2-CH_2-CH_2-CH_3$ 1-heptene

   $CH_3-CH=CH-CH_2-CH_2-CH_2-CH_3$ 2-heptene

   $CH_3-CH_2-CH=CH-CH_2-CH_2-CH_3$ 3-heptene

   Additional choices are shorter-chain alkenes with branches, such as

   $CH_2=\underset{CH_3}{\underset{|}{C}}-CH_2-CH_2-CH_2-CH_3$

   2-methyl-1-hexene

   $CH_2-\underset{CH_3}{\underset{|}{C}}=CH-CH_2-CH_2-CH_3$

   2-methyl-2-hexene

32. Aromatic hydrocarbons have in common the presence of the benzene ring (phenyl group)

33. For benzene, a set of equivalent Lewis structures can be drawn, differing only in the *location* of the three double bonds in the ring. Experimentally, however, benzene does not demonstrate the chemical properties expected for molecules having *any* double bonds. We say that the "extra" electrons that would go into making the second bond of the three double bonds are delocalized around the entire benzene ring: this delocalization of the electrons explains benzene's unique properties.

34. The systematic method for naming monosubstituted benzenes uses the substituent name as a *prefix* for the word benzene. Examples are

chlorobenzene   ethylbenzene

Two monosubsituted benzenes with their own special names are

toluene (methylbenzene)   phenol (hydroxybenzene)

35. *ortho-* refers to adjacent substituents (1,2-); *meta-* refers to two substituents with one unsubstituted carbon atom between them (1,3-); *para-* refers to two substituents with two unsubstituted carbon atoms between them (1,4-).

36. a.

b.

c.

d.

e.

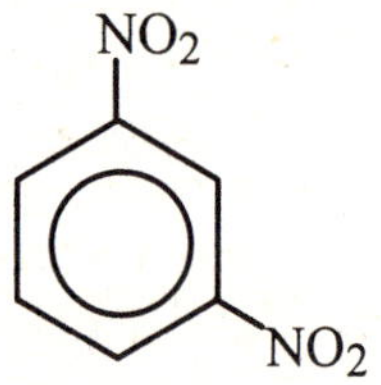

NO₂

NO₂

37.
- a. 1,2-dimethylbenzene (*o*-xylene is its common name)
- b. 1,2,3,4,5,6-hexachlorobenzene
- c. anthracene
- d. 3,5-dichloro-1-methylbenzene (3,5-dichlorotoluene is another name)

38.
- a. alcohol (primary)
- b. ketone
- c. amine
- d. aldehyde

39.
- a. ether
- b. alcohol
- c. alcohol
- d. organic (carboxylic) acid

40. Alcohols are characterized by the presence of the hydroxyl group, –OH. To name an alcohol, the final -*e* is dropped from the name of the parent hydrocarbon, and the ending *ol* is added. A locator number may also be necessary to indicate the location of the hydroxyl group.

41. Primary alcohols have *one* hydrocarbon fragment (alkyl group) bonded to the carbon atom where the –OH group is attached. Secondary alcohols have *two* such alkyl groups attached, and tertiary alcohols contain *three* such alkyl groups. Examples are

ethanol (primary)   CH₃— CH₂ — OH

2-propanol (secondary)

$$CH_3-\underset{\underset{OH}{|}}{CH}-CH_3$$

2-methyl-2-propanol (tertiary)

$$CH_3-\underset{\underset{OH}{|}}{\overset{\overset{CH_3}{|}}{C}}-CH_3$$

42. a. 1-pentanol (primary)
    b. 2-methyl-2-butanol (tertiary)
    c. 3-pentanol (secondary)
    d. 1-propanol (primary)

43. a. $CH_3-CH_2-CH_2-CH_2-CH_2-OH$ primary

   b. $CH_3-\underset{\underset{OH}{|}}{CH}-CH_2-CH_2-CH_3$ secondary

   c. $CH_3-CH_2-\underset{\underset{OH}{|}}{CH}-CH_2-CH_3$ secondary

   d. $CH_3-CH_2-\underset{\underset{CH_3}{|}}{\overset{\overset{OH}{|}}{C}}-CH_2-CH_3$ tertiary

44. Methanol is sometimes called "wood" alcohol, because it formerly was obtained by the heating of wood in the absence of air (this process was called destructive distillation of wood). Currently methanol is most commonly prepared by the catalyzed hydrogenation of carbon monoxide.

$$CO(g) + 2H_2(g) \rightarrow CH_3OH(g)$$

Methanol is an important industrial chemical produced in large amounts every year. It is used as a starting material for the synthesis of acetic acid (CH_3COOH) and of many other important substances. Use has also been made of methanol as a motor fuel for high-performance engines.

45. $C_6H_{12}O_6$–yeast $\rightarrow$ $2CH_3$–CH_2–OH + $2CO_2$

    The yeast necessary for the fermentation process are killed if the concentration of ethanol is over 13%. More concentrated ethanol solutions are most commonly made by distillation.

46. Although much ethanol is produced each year by means of the fermentation process, ethanol is also produced synthetically by hydration of ethene (ethylene)

    $CH_2=CH_2 + H_2O \rightarrow CH_3$–$CH_2OH$

    Ethanol is used in industry as a solvent and as a starting material for the synthesis of more complicated molecules. Mixtures of ethanol and gasoline are used as automobile motor fuels (gasohol).

47. methanol (CH_3OH) – starting material for synthesis of acetic acid and many plastics

    ethylene glycol (CH_2OH–CH_2OH) – automobile antifreeze

    isopropyl alcohol (2-propanol, CH_3–CH(OH)–CH_3) – rubbing alcohol

48. Aldehydes and ketones both contain the *carbonyl* functional group:

    $$\diagdown \!\!\!\! \diagup \text{C} = \text{O}$$

49. Aldehydes and ketones both contain the carbonyl group $\diagdown \!\!\!\! \diagup \text{C} = \text{O}$

    Aldehydes and ketones differ in the *location* of the carbonyl function: aldehydes contain the carbonyl group at the end of a hydrocarbon chain (the carbon atom of the carbonyl group is bonded only to at most one other carbon atom); the carbonyl group of ketones represents one of the interior carbon atoms of a chain (the carbon atom of the carbonyl group is bonded to two other carbon atoms).

50.
    a. 3-hexanone (ethyl propyl ketone)
    b. 2,3-dichlorobutanal (2,3-dichlorobutyraldehyde)
    c. 3,4-dimethylpentanal
    d. 2-methylpropanal
    e. ethyl phenyl ketone

51.
    a.

    $$CH_3-\underset{\underset{O}{\|}}{C}-\phi$$

    [ϕ represents the phenyl group (benzene ring)]

    b.

    $$CH_3-CH_2-CH_2-\underset{\underset{H}{|}}{C}=O$$

c.

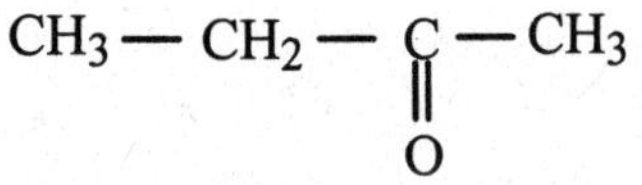

$$CH_3-CH_2-\underset{\underset{O}{\|}}{C}-CH_3$$

d.

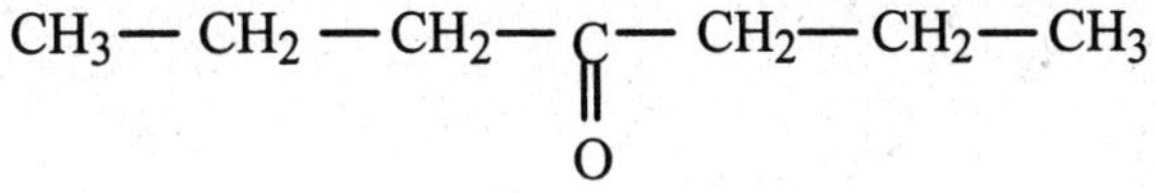

$$CH_3-CH_2-CH_2-\underset{\underset{O}{\|}}{C}-CH_2-CH_2-CH_3$$

e.

$$CH_3-CH_2-CH_2-\underset{\underset{O}{\|}}{C}-CH_2-CH_2-CH_3$$

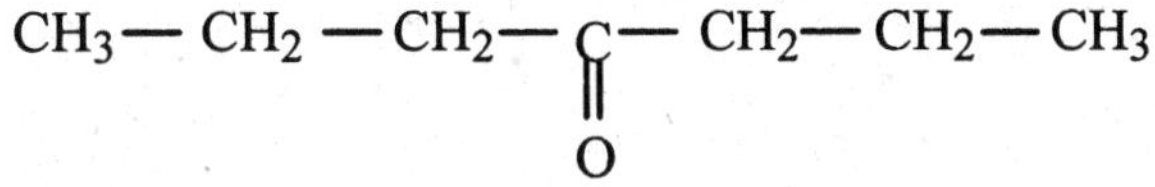

(same molecule as above)

52. Carboxylic acids are typically weak acids.

$$CH_3\text{–}CH_2\text{–}COOH(aq) \rightleftharpoons H^+(aq) + CH_3\text{–}CH_2\text{–}COO^-(aq)$$

53. Acetylsalicylic acid is synthesized from salicylic acid (behaving as an alcohol through its –OH group) and acetic acid.

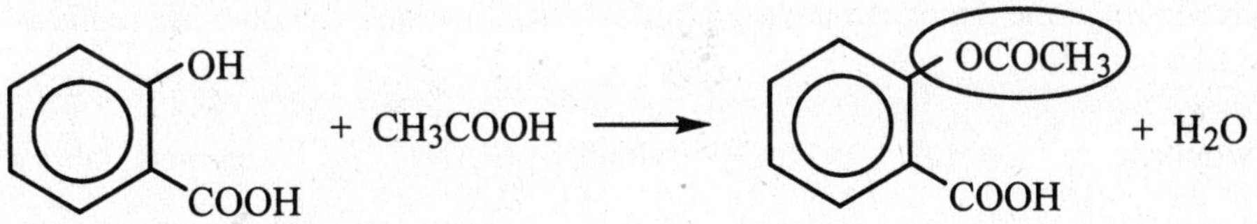

[Structure: salicylic acid (benzene ring with OH and COOH)] + CH₃COOH ⟶ [Structure: benzene ring with OCOCH₃ (circled) and COOH] + H₂O

54. a. 3-methylbutanoic acid
 b. benzoic acid
 c. 2-hydroxypropanoic acid
 d. 3,4-dimethylhexanoic acid

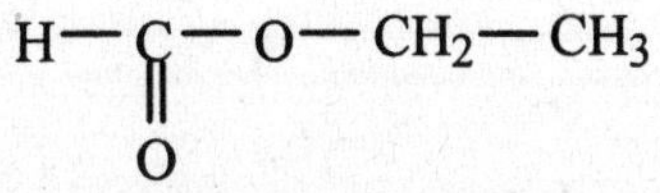

55. a.

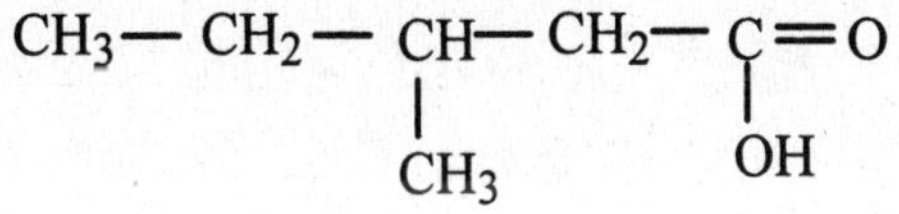

$$CH_3-CH_2-\underset{\underset{CH_3}{|}}{CH}-CH_2-\underset{\underset{OH}{|}}{C}=O$$

b.

$$H-\underset{\underset{O}{\|}}{C}-O-CH_2-CH_3$$

c.

$$\Phi-\underset{\underset{O}{\|}}{C}-O-CH_3$$

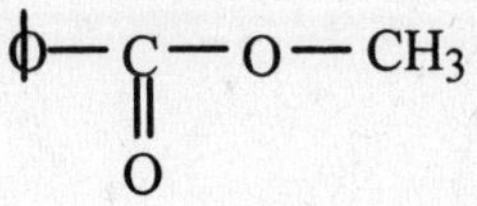

[Φ represents the phenyl group (benzene ring)]

d.

$$CH_3-CH_2-\underset{\underset{Br}{|}}{CH}-\underset{\underset{OH}{|}}{C}=O$$

e.

$$CH_3-CH_2-\underset{\underset{CH_3}{|}}{CH}-\underset{\underset{}{|}}{\overset{\overset{Cl}{|}}{CH}}-\underset{\underset{CH_3}{|}}{CH}-\underset{\underset{}{|}}{\overset{\overset{OH}{|}}{C}}=O$$

56. Polymers are large, usually chain-like molecules, that are built up from smaller molecules. The smaller molecules may combine together and repeat in the chain of the polymer hundreds or thousands of time. The small molecules from which polymers are built up are called *monomers*.

57. A polyester is formed from the reaction of a dialcohol (two –OH groups) with a diacid (two –COOH groups). One –OH group of the alcohol forms an *ester linkage* with one of the –COOH groups of the acid. Since the resulting dimer still possesses an –OH and a –COOH group, the dimer can undergo further esterification reactions. Dacron is a common polyester.

58. acrylonitrile   polyacrylonitrile (PAN)   carpets, fabrics

$$H_2C=\underset{\underset{CN}{|}}{\overset{\overset{H}{|}}{C}} \qquad \left(\!CH_2-\underset{\underset{CN}{|}}{CH}\!\right)_{\!n}$$

butadiene   polybutadiene   tire tread, coating resin

$$H_2C=\overset{\overset{H}{|}}{C}-\overset{\overset{H}{|}}{C}=CH_2 \qquad \left(\!CH_2\,CH=CHCH_2\!\right)_{\!n}$$

59. nylon

$$\left(\!\underset{}{\overset{\overset{H}{|}}{N}}\!\!\left(\!CH_2\!\right)_{\!6}\!\underset{}{\overset{\overset{H}{|}}{N}}\!-\!\overset{\overset{O}{\|}}{C}\!\!\left(\!CH_2\!\right)_{\!4}\!\overset{\overset{O}{\|}}{C}\!\right)$$

dacron

$$\left(\!O-CH_2-CH_2-O-\overset{\overset{O}{\|}}{C}\!-\!\!\bigcirc\!\!-\overset{\overset{O}{\|}}{C}\!\right)$$

Organic Chemistry

60. The general formula is C_nH_{2n+2}. Each succesive alkane differs from the previous or following alkane by CH_2 (sometimes called a methylene unit).

61. The carbon skeletons are

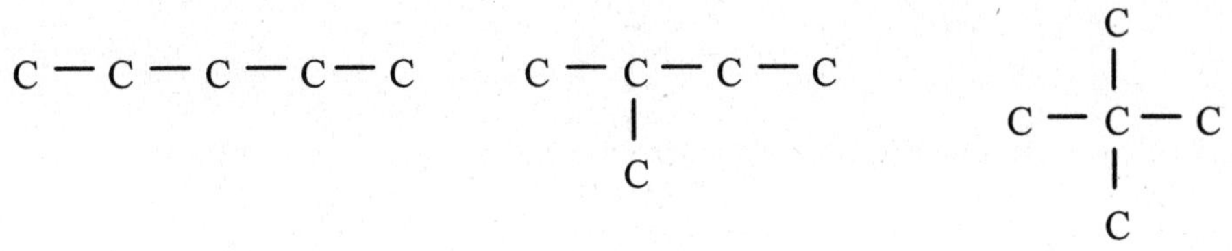

    ```
                                                        C
                                                        |
    C — C — C — C — C      C — C — C — C          C — C — C
                                   |                    |
                                   C                    C
    ```

62. a. 2-chlorobutane

    b. 1,2-dibromoethane

    c. triiodomethane (common name: iodoform)

    d. 2,3,4-trichloropentane

    e. 2,2-dichloro-4-isopropylheptane

63. a.
    $$CH_3-CH(CH_3)-CH(CH_3)-CH_2-CH_2-CH_2-CH_3$$

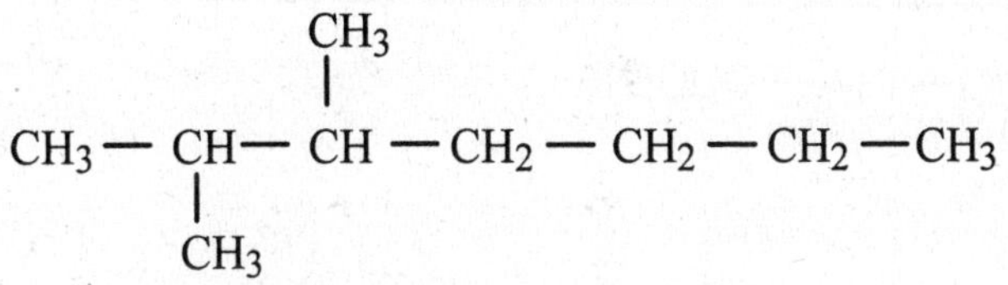

    b.
    $$HO-CH_2-C(CH_3)_2-CHCl-CH_2-CH_2-CH_2-CH_2-CH_3$$
    (with CH₃ and Cl substituents as shown)

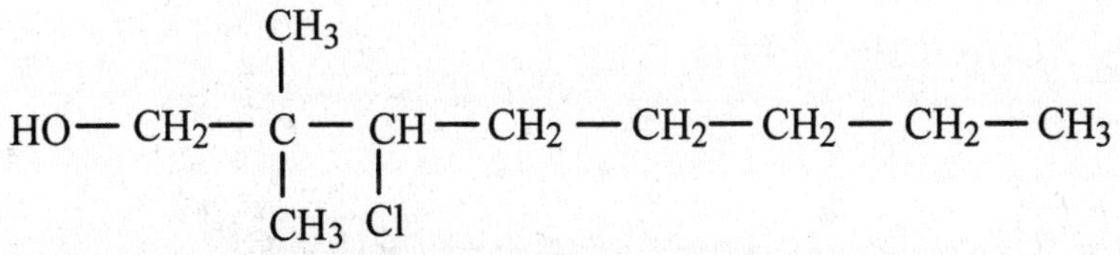

    c.
    $$H_2C=C(Cl)-CH_2-CH_2-CH_2-CH_3$$

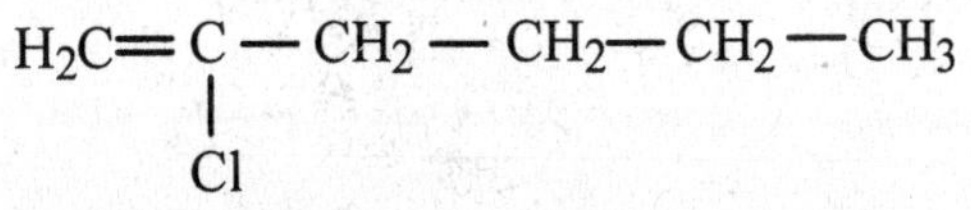

    d.
    $$ClCH_2-CH=CH-CH_2-CH_2-CH_3$$

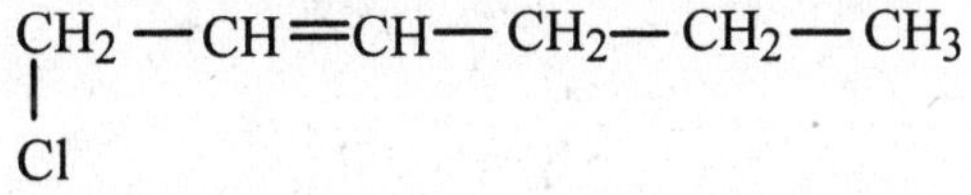

    e. 2-methylphenol (ortho-cresol: benzene ring with OH and CH₃ groups)

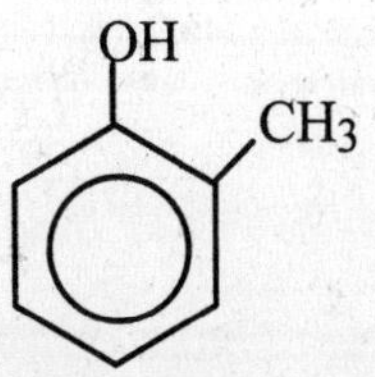

64. a. $CH_3–CH_2Cl$, $CH_2Cl–CH_2Cl$ and various other chlorosubstituted ethanes.

    b. $CH_3CH_2CH_2CH_3$

c.

$$CH_3-CH_2 \atop CH_3-CH_2 \diagdown CBr-CHBr-CH_3$$

65. 1,2,3-trihydroxypropane (1,2,3-propanetriol)

66.

$$\underset{OH}{CH_2}-\underset{OH}{CH}-\underset{OH}{CH}-\underset{OH}{CH}-\underset{OH}{CH}-\underset{H}{C=O}$$

67.

a. $CH_3-CH_2-CH_2-\underset{CH_3}{CH}-\overset{H}{C}=O$

b. $CH_3-\underset{OH}{CH}-CH_2-COOH$

c. $CH_3-\underset{NH_2}{CH}-\overset{H}{C}=O$

d. $CH_3-\underset{O}{\overset{\|}{C}}-CH_2-\underset{O}{\overset{\|}{C}}-CH_2-CH_3$

e.

(benzaldehyde with CH₃ group — 3-methylbenzaldehyde structure)

68.
a. carboxylic acid
b. ketone
c. ester
d. alcohol (phenol)

Chapter 21 Biochemistry

1. Oxygen is the element present in the human body in the largest percentage by mass (65%). Other elements present in the body and there uses are given in Table 21.1.

2. Proteins represent biopolymers of α-amino acids used for many purposes in the human body (structure, enzymes, antibodies, etc.). Proteins make up about 15% of the body by mass.

3. Molar masses of proteins range from a few thousand amu to over 1 million amu. Such molar masses are consistent with proteins being large polymeric molecules.

4. Fibrous proteins provide the structural material of many tissues in the body, and are the chief constituents of hair, cartilage, and muscles. Fibrous proteins consist of lengthwise bundles of polypeptide chains (a fiber). Globular proteins have their polypeptide chains folded into a basically spherical shape and tend to be found in the bloodstream, where they transport and store various needed substances, act as antibodies to fight infections, act as enzymes to catalyze cellular processes, participate in the body's various regulatory systems, and so on.

5. The general structural formula for the alpha amino acids is

$$NH_2-\underset{R}{CH}-\overset{O}{\underset{\|}{C}}-OH$$

   All alpha amino acids contain the carboxyl group (–COOH) and the amino group (–NH$_2$) attached to the α-carbon atom as indicated. In this general formula, R represents the remainder of the amino acid molecule (side-chain): it is this portion of the molecule that differentiates one amino acid from another.

6. The structures of the amino acids are given in detail in Figure 21.2. Generally a side chain is nonpolar if it is mostly hydrocarbon in nature (e.g., alanine, in which the side chain is a methyl group). Side chains are polar if they contain the hydroxyl group (–OH), the sulfhydryl group (–SH), or a second amino (–NH$_2$) or carboxyl (–COOH) group.

7. Since most proteins exist in aqueous media (water) in living creatures, the presence of hydro*phobic* (water-fearing) and hydro*philic* (water-loving) side chains on the amino acids in a protein will greatly influence how that protein's chain interacts with water. In particular, the three-dimensional structure of the protein is greatly influenced by what type of side chains are present in its constituent amino acids.

8. There are six tripeptides possible.

cys–ala–phe	ala–cys–phe	phe–ala–cys
cys–phe–ala	ala–phe–cys	phe–cys–ala

9. a. *ile–ala–gly*

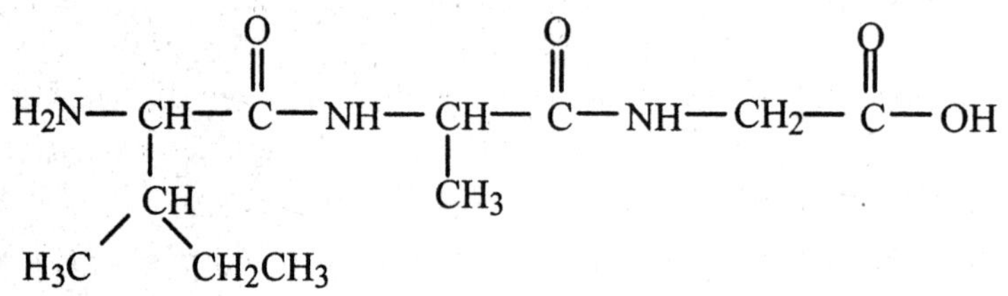

$$H_2N-CH-\underset{\underset{\underset{CH_2CH_3}{|}}{\underset{H_3C}{CH}}}{|}-\overset{O}{\overset{\|}{C}}-NH-\underset{\underset{CH_3}{|}}{CH}-\overset{O}{\overset{\|}{C}}-NH-CH_2-\overset{O}{\overset{\|}{C}}-OH$$

b. *gln–ser*

$$H_2N-\underset{\underset{\underset{\underset{H_2N}{|}}{\underset{C=O}{|}}}{\underset{CH_2}{\underset{|}{CH_2}}}}{CH}-\overset{O}{\overset{\|}{C}}-NH-\underset{\underset{OH}{\underset{|}{CH_2}}}{CH}-\overset{O}{\overset{\|}{C}}-OH$$

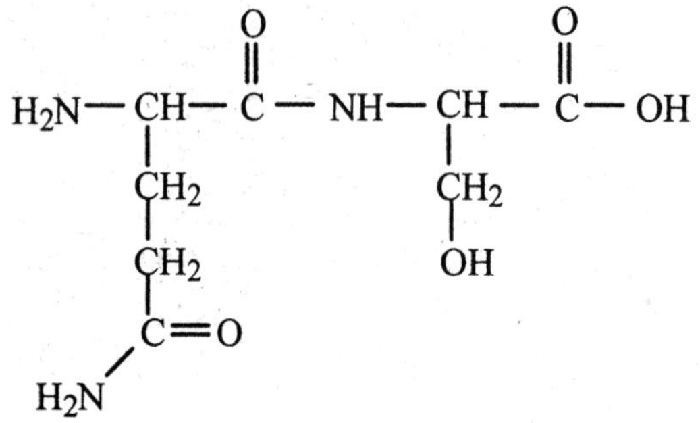

c. *ser–gln*

$$H_2N-\underset{\underset{OH}{\underset{|}{CH_2}}}{CH}-\overset{O}{\overset{\|}{C}}-NH-\underset{\underset{\underset{\underset{H_2N}{|}}{\underset{C=O}{|}}}{\underset{CH_2}{\underset{|}{CH_2}}}}{CH}-\overset{O}{\overset{\|}{C}}-OH$$

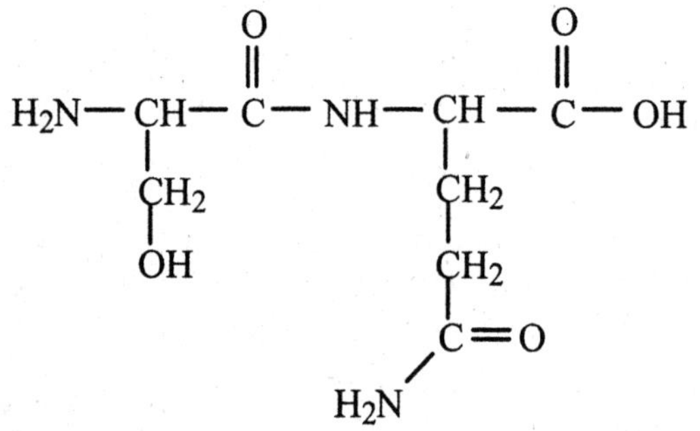

d. *cys–asn–gly*

$$H_2N-\underset{\underset{SH}{\underset{|}{CH_2}}}{CH}-\overset{O}{\overset{\|}{C}}-NH-\underset{\underset{\underset{H_2N}{|}}{\underset{C=O}{|}}\underset{CH_2}{|}}{CH}-\overset{O}{\overset{\|}{C}}-NH-CH_2-\overset{O}{\overset{\|}{C}}-OH$$

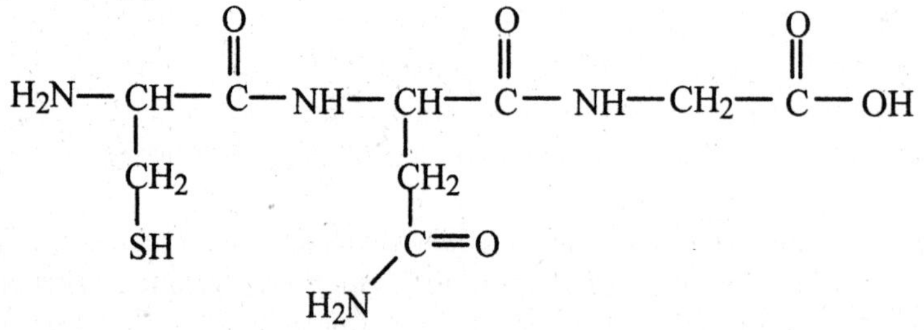

10. *phe–ala–gly*

$$H_2N-\underset{\underset{\text{N terminal}}{}}{\underset{\underset{\bigcirc}{\underset{|}{CH_2}}}{CH}}\!-\!\boxed{\overset{O}{\overset{\|}{C}}-NH}\!-\!\underset{\underset{CH_3}{|}}{CH}\!-\!\boxed{\overset{O}{\overset{\|}{C}}-NH}\!-\!CH_2-\underset{\underset{\text{C terminal}}{}}{\overset{O}{\overset{\|}{C}}-OH}$$

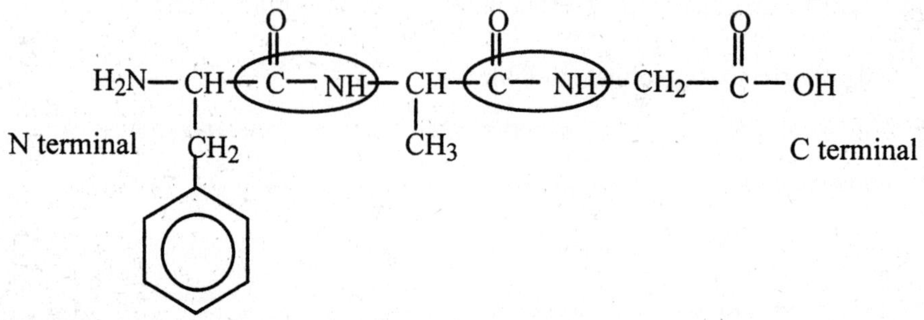

phe–gly–ala

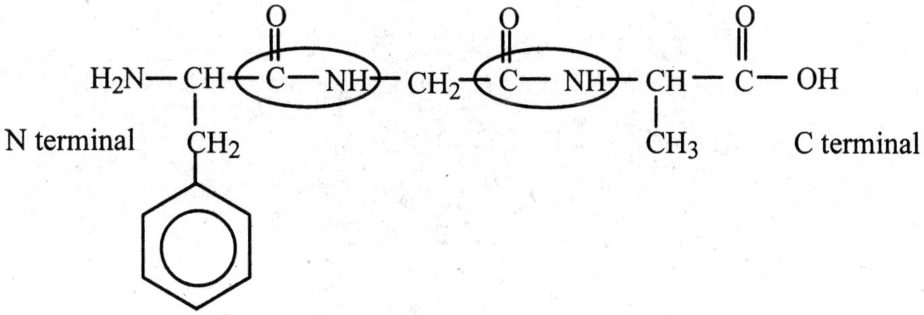
N terminal
C terminal

ala–phe–gly

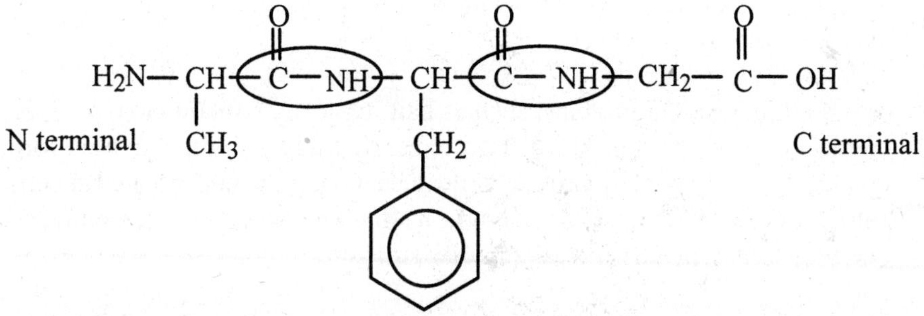
N terminal
C terminal

ala–gly–phe

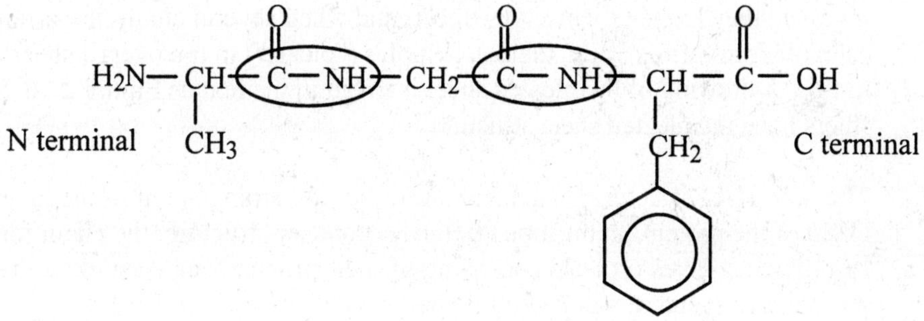
N terminal
C terminal

gly–phe–ala

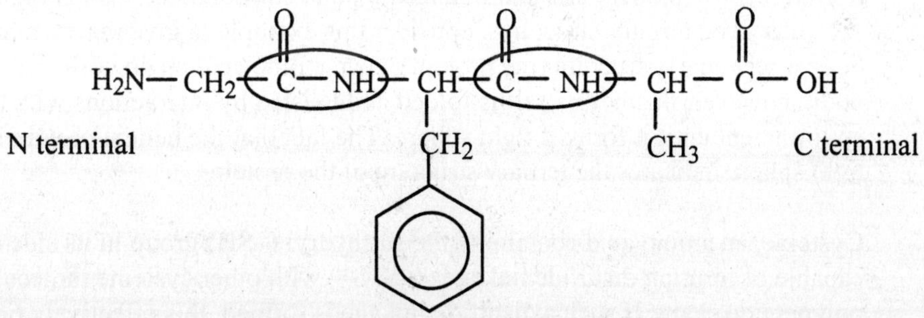
N terminal
C terminal

gly–ala–phe

$H_2N-CH_2-\underbrace{C-NH}_{}-CH-\underbrace{C-NH}_{}-CH-C-OH$

with C=O above each C, CH₃ below the second CH, and CH₂–phenyl below the third CH.

N terminal   CH₃   CH₂   C terminal

11. The secondary structure of a protein describes, in general, the arrangement in space of the protein's polypeptide chain. The most common secondary structures are the alpha-helix and the beta-pleated sheet.

12. Long, thin, resilient proteins, such as hair, typically contain elongated, elastic alpha-helical protein molecules. Other proteins, such as silk, which in bulk form sheets or plates, typically contain protein molecules having the beta pleated sheet secondary structure. Proteins which do not have a structural function in the body, such as hemoglobin, typically have a globular structure.

13. In the pleated sheet secondary structure, a large number of similar polypeptide chains are arranged lengthwise next to each other to form a wide sheet of protein. Because the individual polypeptide chains contain the normal bond angles associated with the atoms involved they are not themselves linear, and when several chains are arranged next to each other to form a sheet, there are ripples ("pleats") in the overall sheet of protein due to this. A drawing of the pleated sheet is given in the text as Figure 21.6. Silk and muscle fibers have the pleated sheet structure

14. The secondary structure, in general, describes the arrangement of the long polypeptide chain of the protein. In the alpha-helical secondary structure, the chain forms a coil or spiral, which gives proteins consisting of such structures an elasticity or resilience. Such proteins are found in wool, hair, and tendons.

15. The tertiary structure of a protein represents its specific, overall shape when it occurs in its natural environment and is influenced by that environment. To distinguish between the secondary and tertiary structures, consider this example: a given protein has an alpha helical secondary structure (the protein's own amino acid chain coils in a helix); in the body, however, this helix itself is folded and twisted by interactions with the protein's environment until it forms a tight sphere. The fact that the helical protein is folded into a tight sphere indicates the tertiary structure of the protein.

16. Cysteine, an amino acid containing the sulfhydryl (–SH) group in its side chain, is capable of forming disulfide linkages (–S–S–) with other cysteine molecules in the same polypeptide chain. If such a disulfide linkage is formed, this effectively ties together two portions of the polypeptide, producing a kink or knot in the chain, which leads in part to the protein's overall 3-dimensional shape (tertiary structure). Cysteine, and the disulfide linkages it forms, is responsible for the curling of hair (whether naturally or by a permanent wave).

17. Denaturation of a protein represents the breaking down of the protein's tertiary structure. If the environment of a protein is changed from the normal environment of the protein, the specific folding and twisting of the protein's polypeptide chain will change to accommodate the new environment. If the protein's tertiary structure is changed, the protein most likely will no longer have whatever function in the body it ordinarily possesses. Cooking of an egg (adding heat to the environment of the protein) causes the protein albumin in the white of the egg to coagulate. Adding heavy metal ions (lead or mercury, for example) can disrupt the inter-chain linkages that contribute to the protein's tertiary structure. A permanent hair wave works by deliberately changing the hair protein's structure.

18. Collagen has an alpha-helical secondary structure. Collagen's function in the body is as the raw material of which tendons are constructed. The long, springy structure of the alpha-helix is responsible for collagen's strength and elasticity.

19. Proteins that catalyze biochemical processes are called *enzymes*.

20. Antibodies are special proteins which are synthesized in the body in response to foreign substances such as bacteria or viruses. Although there are usually specific antibodies for specific invaders, the antibody interferon offers general protection against invasion by viruses.

21. Antibodies are special proteins that are synthesized in the body in response to foreign substances such as bacteria. Interferon is a protein that helps to protect cells against viral infection. The process of blood-clotting involves several proteins.

22. In a permanent wave, cross-linkages between adjacent polypeptide chains of the protein are broken chemically, and then reformed chemically in a new location. The primary cross-linkage involved is a disulfide linkage between cysteine units in the polypeptide chains. It is primarily the tertiary structure of the hair protein which is affected by a permanent wave, although if the waving lotion is left on the hair too long, the secondary structure can also be affected (making the hair very "frizzy").

23. Enzymes are typically 1 to 10 *million* times more efficient than inorganic catalysts. Enzymes are much more efficient than any inorganic catalyst.

24. The molecule acted upon by an enzyme is referred to as the enzyme's substrate. When we say that an enzyme is specific for a particular substrate, we mean that the enzyme will catalyze the reactions of that molecule and that molecule only.

25. The action of an enzyme on its substrate takes place at a specific portion of the polypeptide chain called the *active site*.

26. Figure 21.11 illustrates the lock and key model clearly. The lock-and-key model for enzyme action postulates that the structures of the enzyme and its substrate are complementary, such that the active site of the enzyme and the portion of the substrate molecule to be acted upon can fit together very closely. The structures of these portions of the molecules are unique to the particular enzyme-substrate pair, and they fit together much like a particular key is necessary to work in a given lock.

27. Simple sugars typically contain several hydroxyl (–OH) groups, as well as the carbonyl (C=O) function (making the sugar either an aldehyde or a ketone, as well as a polyalcohol).

28. Sugars contain an aldehyde or ketone functional group (carbonyl group), as well as several –OH groups (hydroxyl groups).

29. In solution, functional groups at opposite ends of the simple sugar molecules react with each other, forming the sugar into a *ring* or *cyclic* structure (see Figure 21.12 in the text for the structure).

30. a. glucose

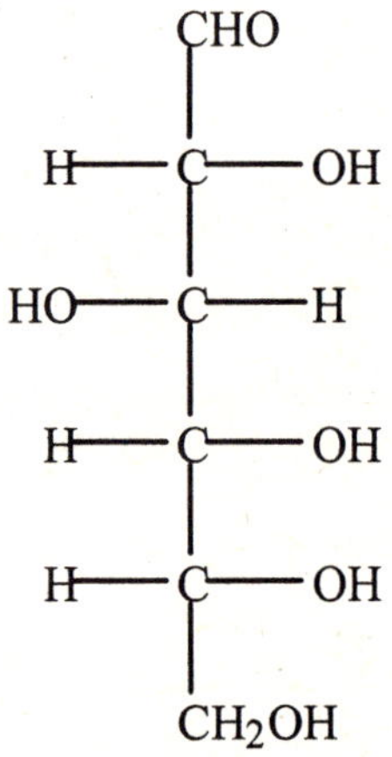

$$\begin{array}{c} CHO \\ H-C-OH \\ HO-C-H \\ H-C-OH \\ H-C-OH \\ CH_2OH \end{array}$$

b. ribose

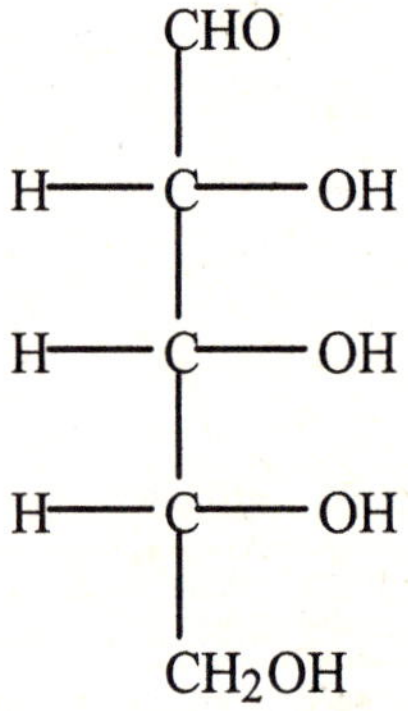

$$\begin{array}{c} CHO \\ H-C-OH \\ H-C-OH \\ H-C-OH \\ CH_2OH \end{array}$$

c. ribulose

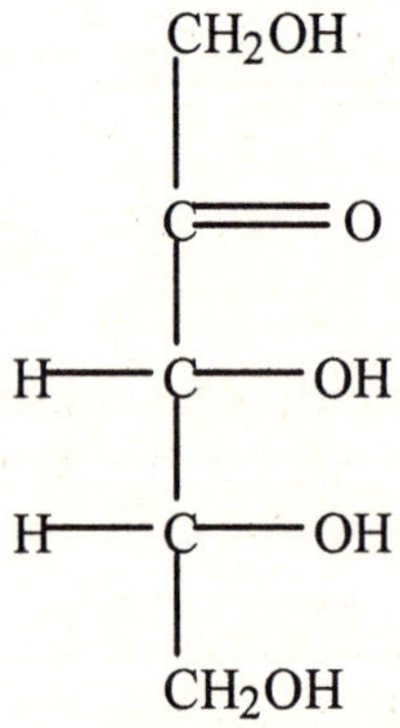

$$\begin{array}{c} CH_2OH \\ C=O \\ H-C-OH \\ H-C-OH \\ CH_2OH \end{array}$$

d. galactose

```
        CHO
         |
    H —— C —— OH
         |
   HO —— C —— H
         |
   HO —— C —— H
         |
    H —— C —— OH
         |
        CH₂OH
```

31. Deoxyribonucleic acid (DNA) is the molecule responsible for coding and storing genetic information in the cell, and for subsequently transmitting that information. DNA is found in the nucleus of each cell. The molar mass of DNA depends on the complexity of the plant or animal species involved, but human DNA may have a molar mass as large as two billion grams.

32. Uracil (RNA only); cytosine (DNA, RNA); thymine (DNA only); adenine (DNA, RNA); guanine (DNA, RNA)

33. In a strand of DNA, the phosphate group and the sugar molecule of adjacent nucleotides become bonded to each other. The chain-portion of the DNA molecule, therefore, consists of alternating phosphate groups and sugar molecules. The nitrogen bases are found sticking out from the side of this phosphate-sugar chain, being bonded to the sugar molecules.

34. When the two strands of a DNA molecule are compared, it is found that a given base in one strand is always found paired with a particular base in the other strand. Because of the shapes and side atoms along the rings of the nitrogen bases, only certain pairs are able to approach and hydrogen-bond with each other in the double helix. Adenine is always found paired with thymine; cytosine is always found paired with guanine. When a DNA helix unwinds for replication during cell division, only the appropriate complementary bases are able to approach and bond to the nitrogen bases of each strand. For example, for a guanine-cytosine pair in the original DNA, when the two strands separate, only a new cytosine molecule can approach and bond to the original guanine, and only a new guanine molecule can approach and bond to the original cytosine.

35. A given section of the DNA molecule called a *gene* contains the specific information necessary for construction of a particular protein.

36. Messenger RNA molecules are synthesized to be complementary to a portion (gene) of the DNA molecule in the cell, and serve as the template or pattern upon which a protein will be constructed (a particular group of nitrogen bases on *m*-RNA is able to accommodate and specify a particular amino acid in a particular location in the protein). Transfer RNA molecules are much smaller than *m*-RNA, and their structure accommodates only a single specific amino acid molecule: transfer RNA molecules

"find" their specific amino acid in the cellular fluids, and bring it to *m*-RNA where it is added to the protein molecule being synthesized.

37. Rather than having some common structural feature, substances are classified as lipids based on their solubility characteristics. Lipids are water-insoluble substances that can be extracted from cells by nonpolar organic solvents such as benzene or carbon tetrachloride.

38. Saturated: butyric acid, caproic acid, lauric acid, stearic acid

    Unsaturated: oleic acid, linoleic acid, linolenic acid

    Most unsaturated fatty acids are derived from vegetable matter (plants).

39. Saponification is the production of a *soap* by treatment of a triglyceride with a strong base such as NaOH.

    triglyceride + 3NaOH → glycerol + 3Na$^+$soap$^-$

    $$\begin{array}{c} CH_2-O-\overset{O}{\underset{\|}{C}}-R \\ | \\ CH_2-O-\overset{O}{\underset{\|}{C}}-R' \\ | \\ CH_2-O-\overset{O}{\underset{\|}{C}}-R'' \end{array} + 3NaOH \rightarrow \begin{array}{c} CH_2-OH \\ | \\ CH_2-OH \\ | \\ CH_2-OH \end{array} + \begin{array}{c} RCOONa \\ R'COONa \\ R''COONa \end{array}$$

40. Soaps have both a nonpolar nature (due to the long chain of the fatty acid) and an ionic nature (due to the charge on the carboxyl group). In water, soap anions form aggregates called micelles, in which the water-repelling hydrocarbon chains are oriented towards the interior of the aggregate, with the ionic, water-attracting carboxyl groups oriented towards the outside. Most dirt has a greasy nature. A soap micelle is able to interact with a grease molecule, pulling the grease molecule into the hydrocarbon interior of the micelle. When the clothing is rinsed, the micelle containing the grease is washed away. See Figures 21.22 and 21.23.

41. Cholesterol is the naturally occurring steroid from which the body synthesizes other needed steroids. Since cholesterol is insoluble in water, it is thought that having too large a concentration of this substance in the bloodstream may lead to its deposition and build up on the walls of blood vessels, causing their eventual blockage.

42. v 43. i
44. t 45. m
46. x 47. u
48. q 49. f
50. k 51. g
52. y 53. r

Biochemistry

54.	c	55.	p
56.	n	57.	o
58.	e	59.	b
60.	s	61.	d
62.	h	63.	a